Student Edition

Eureka Math
Grade 8
Modules 3, 4, & 5

Special thanks go to the Gordon A. Cain Center and to the Department of Mathematics at Louisiana State University for their support in the development of *Eureka Math*.

For a free *Eureka Math* Teacher Resource Pack, Parent Tip Sheets, and more please visit www.Eureka.tools

Published by the non-profit Great Minds

Copyright © 2015 Great Minds. No part of this work may be reproduced, sold, or commercialized, in whole or in part, without written permission from Great Minds. Non-commercial use is licensed pursuant to a Creative Commons Attribution-NonCommercial-ShareAlike 4.0 license; for more information, go to http://greatminds.net/maps/math/copyright. "Great Minds" and "Eureka Math" are registered trademarks of Great Minds.

Printed in the U.S.A.

This book may be purchased from the publisher at eureka-math.org

1 2 3 4 5 6 7 8 BAB 25 24 23 22 21

ISBN 978-1-63255-321-8

Lesson 1: What Lies Behind "Same Shape"?

Classwork

Exploratory Challenge

Two geometric figures are said to be similar if they have the same shape but not necessarily the same size. Using that informal definition, are the following pairs of figures similar to one another? Explain.

Pair A:

similar

Pair B:

different

Pair C:

same

Pair D:

similar

Pair E:  different

Pair F:  similar

Pair G:  different

Pair H:

Exercises

1. Given $|OP| = 5$ in.

   a. If segment OP is dilated by a scale factor $r = 4$, what is the length of segment OP'?

   b. If segment OP is dilated by a scale factor $r = \frac{1}{2}$, what is the length of segment OP'?

EUREKA MATH™

Use the diagram below to answer Exercises 2–6. Let there be a dilation from center O. Then, $Dilation(P) = P'$ and $Dilation(Q) = Q'$. In the diagram below, $|OP| = 3$ cm and $|OQ| = 4$ cm, as shown.

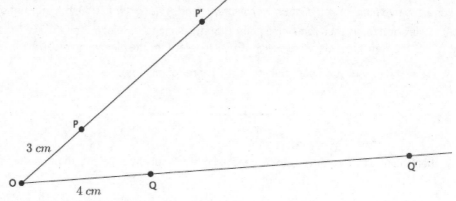

2. If the scale factor is $r = 3$, what is the length of segment OP'?

3. Use the definition of dilation to show that your answer to Exercise 2 is correct.

4. If the scale factor is $r = 3$, what is the length of segment OQ'?

5. Use the definition of dilation to show that your answer to Exercise 4 is correct.

6. If you know that $|OP| = 3$, $|OP'| = 9$, how could you use that information to determine the scale factor?

Lesson Summary

Definition: For a positive number r, a *dilation with center O and scale factor r* is the transformation of the plane that maps O to itself, and maps each remaining point P of the plane to its image P' on the ray $\overrightarrow{OP}$ so that $|OP'| = r|OP|$. That is, it is the transformation that assigns to each point P of the plane a point $Dilation(P)$ so that

1. $Dilation(O) = O$ (i.e., a dilation does not move the center of dilation).

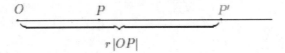

2. If $P \neq O$, then the point $Dilation(P)$ (to be denoted more simply by P') is the point on the ray $\overrightarrow{OP}$ so that $|OP'| = r|OP|$.

In other words, a dilation is a rule that moves each point P along the ray emanating from the center O to a new point P' on that ray such that the distance $|OP'|$ is r times the distance $|OP|$.

Problem Set

1. Let there be a dilation from center O. Then, $Dilation(P) = P'$ and $Dilation(Q) = Q'$. Examine the drawing below. What can you determine about the scale factor of the dilation?

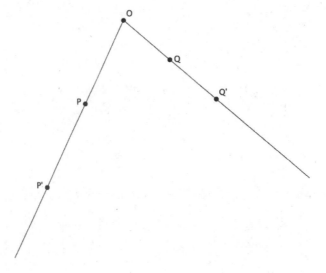

Lesson 1:   What Lies Behind "Same Shape"?

EUREKA
MATH™

2.   Let there be a dilation from center O. Then, $Dilation(P) = P'$, and $Dilation(Q) = Q'$. Examine the drawing below. What can you determine about the scale factor of the dilation?

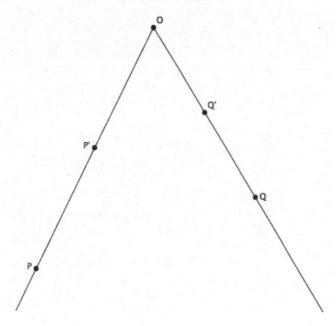

3.   Let there be a dilation from center O with a scale factor $r = 4$. Then, $Dilation(P) = P'$ and $Dilation(Q) = Q'$. $|OP| = 3.2$ cm, and $|OQ| = 2.7$ cm, as shown. Use the drawing below to answer parts (a) and (b). The drawing is not to scale.

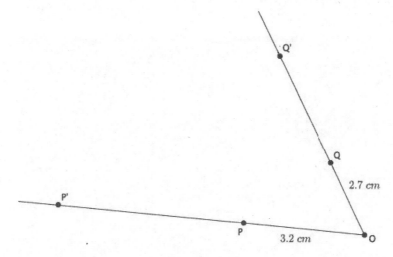

a.   Use the definition of dilation to determine $|OP'|$.

b.   Use the definition of dilation to determine $|OQ'|$.

4. Let there be a dilation from center O with a scale factor r. Then, $Dilation(A) = A'$, $Dilation(B) = B'$, and $Dilation(C) = C'$. $|OA| = 3$, $|OB| = 15$, $|OC| = 6$, and $|OB'| = 5$, as shown. Use the drawing below to answer parts (a)–(c).

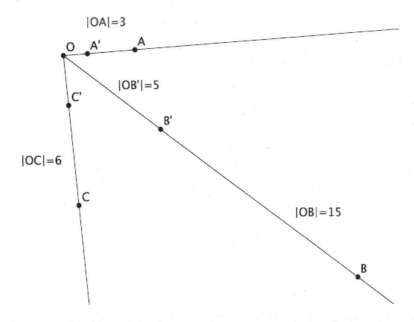

a. Using the definition of dilation with lengths OB and OB', determine the scale factor of the dilation.

b. Use the definition of dilation to determine $|OA'|$.

c. Use the definition of dilation to determine $|OC'|$.

Lesson 1: What Lies Behind "Same Shape"?

© 2015 Great Minds. eureka-math.org
G8-M3M4M5-SE-B2-1.3.1-01.2016

EUREKA MATH™

Lesson 2: Properties of Dilations

Classwork

Examples 1–2: Dilations Map Lines to Lines

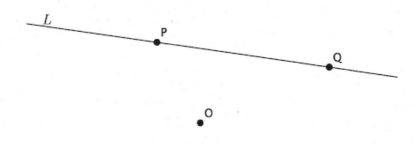

Example 3: Dilations Map Lines to Lines

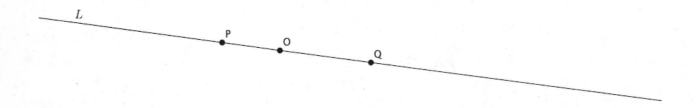

Exercise

Given center O and triangle ABC, dilate the triangle from center O with a scale factor $r = 3$.

a.   Note that the triangle ABC is made up of segments AB, BC, and CA. Were the dilated images of these segments still segments?

EUREKA
MATH™

b. Measure the length of the segments AB and $A'B'$. What do you notice? (Think about the definition of dilation.)

c. Verify the claim you made in part (b) by measuring and comparing the lengths of segments BC and $B'C'$ and segments CA and $C'A'$. What does this mean in terms of the segments formed between dilated points?

d. Measure $\angle ABC$ and $\angle A'B'C'$. What do you notice?

e. Verify the claim you made in part (d) by measuring and comparing the following sets of angles: (1) $\angle BCA$ and $\angle B'C'A'$ and (2) $\angle CAB$ and $\angle C'A'B'$. What does that mean in terms of dilations with respect to angles and their degrees?

Lesson Summary

Dilations map lines to lines, rays to rays, and segments to segments. Dilations map angles to angles of the same degree.

Problem Set

1. Use a ruler to dilate the following figure from center O, with scale factor $r = \frac{1}{2}$.

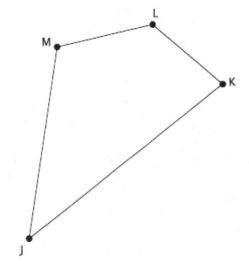

Lesson 2:   Properties of Dilations

2.   Use a compass to dilate the figure $ABCDE$ from center O, with scale factor $r = 2$.

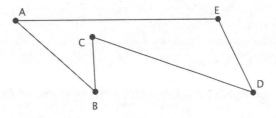

a.   Dilate the same figure, $ABCDE$, from a new center, O', with scale factor $r = 2$. Use double primes ($A''B''C''D''E''$) to distinguish this image from the original.

b.   What rigid motion, or sequence of rigid motions, would map $A''B''C''D''E''$ to $A'B'C'D'E'$?

Lesson 2:   Properties of Dilations

S.11

EUREKA
MATH™

© 2015 Great Minds. eureka-math.org
G8-M3M4M5-SE-B2-1.3.1-01.2016

3. Given center O and triangle ABC, dilate the figure from center O by a scale factor of $r = \frac{1}{4}$. Label the dilated triangle $A'B'C'$.

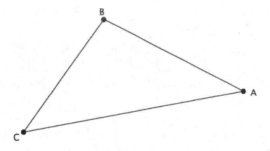

         O

4. A line segment AB undergoes a dilation. Based on today's lesson, what is the image of the segment?

5. $\angle GHI$ measures 78°. After a dilation, what is the measure of $\angle G'H'I'$? How do you know?

EUREKA
MATH™

Lesson 3: Examples of Dilations

Classwork

Example 1

Dilate circle A from center O at the origin by scale factor $r = 3$.

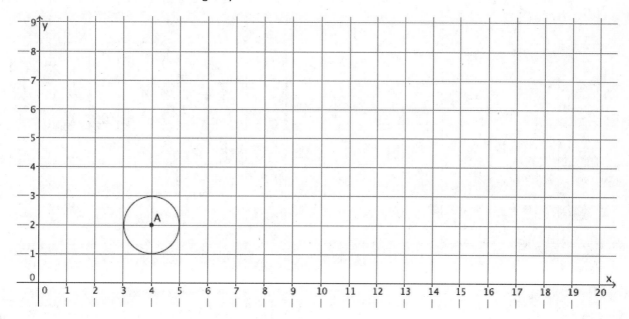

Exercises 1–2

1. Dilate ellipse E, from center O at the origin of the graph, with scale factor $r = 2$. Use as many points as necessary to develop the dilated image of ellipse E.

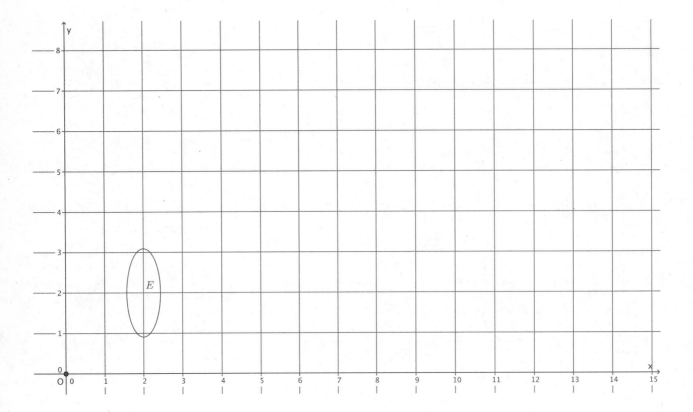

2. What shape was the dilated image?

EUREKA
MATH

Exercise 3

3. Triangle ABC has been dilated from center O by a scale factor of $r = \frac{1}{4}$ denoted by triangle $A'B'C'$. Using a centimeter ruler, verify that it would take a scale factor of $r = 4$ from center O to map triangle $A'B'C'$ onto triangle ABC.

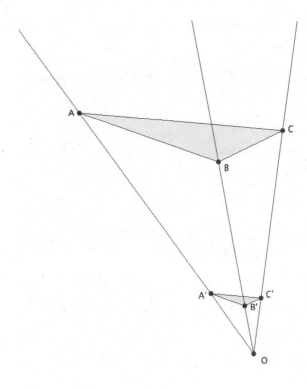

EUREKA
MATH™

Lesson Summary

Dilations map circles to circles and ellipses to ellipses.

If a figure is dilated by scale factor r, we must dilate it by a scale factor of $\frac{1}{r}$ to bring the dilated figure back to the original size. For example, if a scale factor is $r = 4$, then to bring a dilated figure back to the original size, we must dilate it by a scale factor $r = \frac{1}{4}$.

Problem Set

1. Dilate the figure from center O by a scale factor $r = 2$. Make sure to use enough points to make a good image of the original figure.

2. Describe the process for selecting points when dilating a curved figure.

3. A figure was dilated from center O by a scale factor of $r = 5$. What scale factor would shrink the dilated figure back to the original size?

4. A figure has been dilated from center O by a scale factor of $r = \frac{7}{6}$. What scale factor would shrink the dilated figure back to the original size?

5. A figure has been dilated from center O by a scale factor of $r = \frac{3}{10}$. What scale factor would magnify the dilated figure back to the original size?

EUREKA
MATH™

Lesson 4: Fundamental Theorem of Similarity (FTS)

Classwork

Exercise

In the diagram below, points R and S have been dilated from center O by a scale factor of $r = 3$.

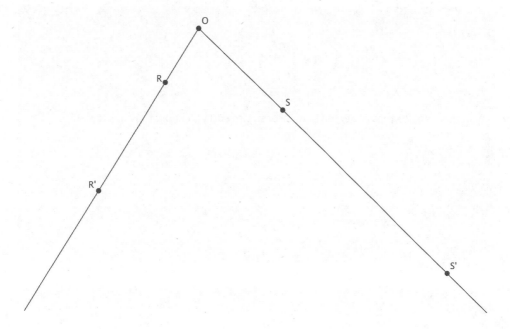

a. If $|OR| = 2.3$ cm, what is $|OR'|$?

b. If $|OS| = 3.5$ cm, what is $|OS'|$?

EUREKA
MATH™

Lesson 4: Fundamental Theorem of Similarity (FTS)

S.17

© 2015 Great Minds. eureka-math.org
G8-M3M4M5-SE-B2-1.3.1-01.2016

c.   Connect the point R to the point S and the point R' to the point S'. What do you know about the lines that contain segments RS and $R'S'$?

d.   What is the relationship between the length of segment RS and the length of segment $R'S'$?

e.   Identify pairs of angles that are equal in measure. How do you know they are equal?

EUREKA
MATH™

Lesson Summary

THEOREM: Given a dilation with center O and scale factor r, then for any two points P and Q in the plane so that O, P, and Q are not collinear, the lines PQ and $P'Q'$ are parallel, where $P' = Dilation(P)$ and $Q' = Dilation(Q)$, and furthermore, $|P'Q'| = r|PQ|$.

Problem Set

1. Use a piece of notebook paper to verify the fundamental theorem of similarity for a scale factor r that is $0 < r < 1$.

   ✓ Mark a point O on the first line of notebook paper.

   ✓ Mark the point P on a line several lines down from the center O. Draw a ray, $\overrightarrow{OP}$. Mark the point P' on the ray and on a line of the notebook paper closer to O than you placed point P. This ensures that you have a scale factor that is $0 < r < 1$. Write your scale factor at the top of the notebook paper.

   ✓ Draw another ray, $\overrightarrow{OQ}$, and mark the points Q and Q' according to your scale factor.

   ✓ Connect points P and Q. Then, connect points P' and Q'.

   ✓ Place a point, A, on the line containing segment PQ between points P and Q. Draw ray $\overrightarrow{OA}$. Mark point A' at the intersection of the line containing segment $P'Q'$ and ray $\overrightarrow{OA}$.

   a. Are the lines containing segments PQ and $P'Q'$ parallel lines? How do you know?

   b. Which, if any, of the following pairs of angles are equal in measure? Explain.

      i. $\angle OPQ$ and $\angle OP'Q'$

      ii. $\angle OAQ$ and $\angle OA'Q'$

      iii. $\angle OAP$ and $\angle OA'P'$

      iv. $\angle OQP$ and $\angle OQ'P'$

   c. Which, if any, of the following statements are true? Show your work to verify or dispute each statement.

      i. $|OP'| = r|OP|$

      ii. $|OQ'| = r|OQ|$

      iii. $|P'A'| = r|PA|$

      iv. $|A'Q'| = r|AQ|$

   d. Do you believe that the fundamental theorem of similarity (FTS) is true even when the scale factor is $0 < r < 1$? Explain.

2. Caleb sketched the following diagram on graph paper. He dilated points B and C from center O.

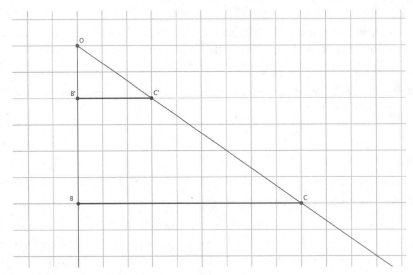

a. What is the scale factor r? Show your work.

b. Verify the scale factor with a different set of segments.

c. Which segments are parallel? How do you know?

d. Which angles are equal in measure? How do you know?

3. Points B and C were dilated from center O.

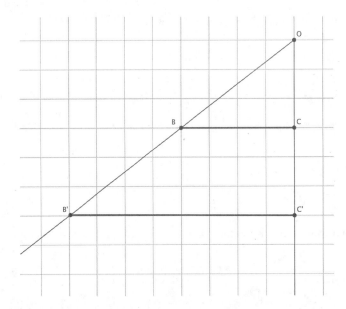

a. What is the scale factor r? Show your work.

b. If $|OB| = 5$, what is $|OB'|$?

c. How does the perimeter of triangle OBC compare to the perimeter of triangle $OB'C'$?

d. Did the perimeter of triangle $OB'C' = r \times$ (perimeter of triangle OBC)? Explain.

Lesson 4: Fundamental Theorem of Similarity (FTS)

EUREKA MATH™

Lesson 5: First Consequences of FTS

Exercise 1

In the diagram below, points P and Q have been dilated from center O by scale factor r. $\overline{PQ} \parallel \overline{P'Q'}$, $|PQ| = 5$ cm, and $|P'Q'| = 10$ cm.

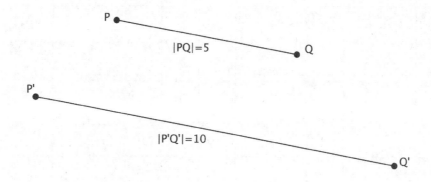

a.   Determine the scale factor r.

b.   Locate the center O of dilation. Measure the segments to verify that $|OP'| = r|OP|$ and $|OQ'| = r|OQ|$. Show your work below.

Exercise 2

In the diagram below, you are given center O and ray $\overrightarrow{OA}$. Point A is dilated by a scale factor $r = 4$. Use what you know about FTS to find the location of point A'.

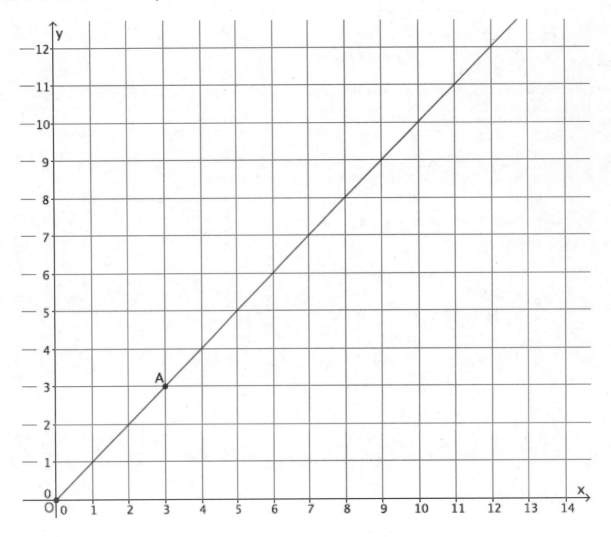

EUREKA
MATH™

Exercise 3

In the diagram below, you are given center O and ray $\overrightarrow{OA}$. Point A is dilated by a scale factor $r = \dfrac{5}{12}$. Use what you know about FTS to find the location of point A'.

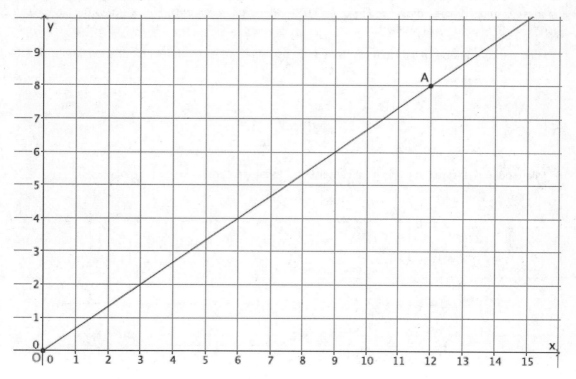

EUREKA
MATH™

Lesson 5:   First Consequences of FTS

S.23

© 2015 Great Minds. eureka-math.org
G8-M3M4M5-SE-B2-1.3.1-01.2016

Lesson Summary

Converse of the fundamental theorem of similarity:

If lines PQ and $P'Q'$ are parallel and $|P'Q'| = r|PQ|$, then from a center O, $P' = Dilation(P)$, $Q' = Dilation(Q)$, $|OP'| = r|OP|$, and $|OQ'| = r|OQ|$.

To find the coordinates of a dilated point, we must use what we know about FTS, dilation, and scale factor.

Problem Set

1. Dilate point A, located at $(3, 4)$ from center O, by a scale factor $r = \dfrac{5}{3}$.

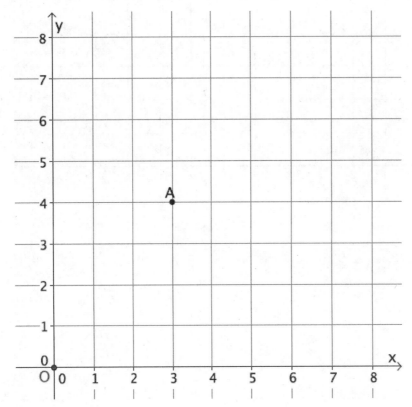

    What is the precise location of point A'?

EUREKA
MATH™

2. Dilate point A, located at $(9, 7)$ from center O, by a scale factor $r = \frac{4}{9}$. Then, dilate point B, located at $(9, 5)$ from center O, by a scale factor of $r = \frac{4}{9}$. What are the coordinates of points A' and B'? Explain.

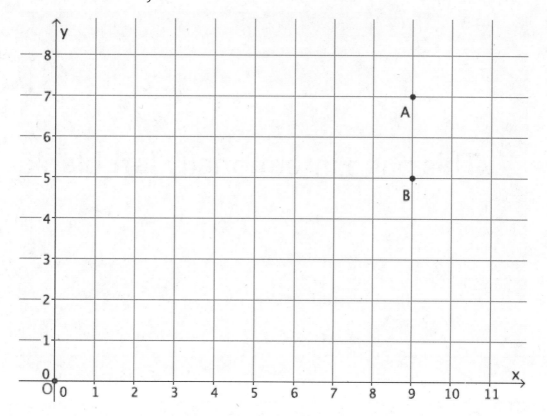

3. Explain how you used the fundamental theorem of similarity in Problems 1 and 2.

This page intentionally left blank

Lesson 6: Dilations on the Coordinate Plane

Classwork

Exercises 1–5

1. Point $A(7, 9)$ is dilated from the origin by scale factor $r = 6$. What are the coordinates of point A'?

2. Point $B(-8, 5)$ is dilated from the origin by scale factor $r = \frac{1}{2}$. What are the coordinates of point B'?

3. Point $C(6, -2)$ is dilated from the origin by scale factor $r = \frac{3}{4}$. What are the coordinates of point C'?

4. Point $D(0, 11)$ is dilated from the origin by scale factor $r = 4$. What are the coordinates of point D'?

5. Point $E(-2, -5)$ is dilated from the origin by scale factor $r = \frac{3}{2}$. What are the coordinates of point E'?

EUREKA
MATH™

Lesson 6:   Dilations on the Coordinate Plane

© 2015 Great Minds. eureka-math.org
G8-M3M4M5-SE-B2-1.3.1-01.2016

S.27

Exercises 6–8

6. The coordinates of triangle ABC are shown on the coordinate plane below. The triangle is dilated from the origin by scale factor $r = 12$. Identify the coordinates of the dilated triangle $A'B'C'$.

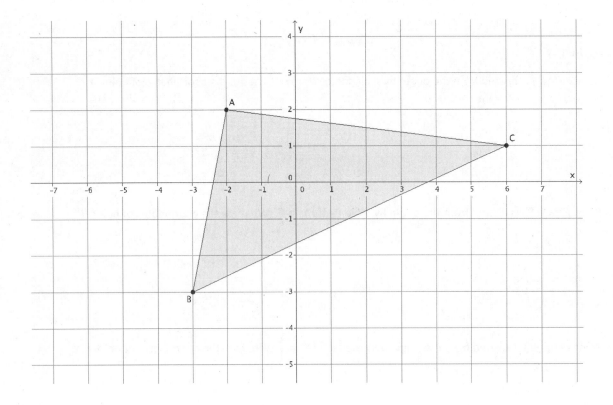

Lesson 6:   Dilations on the Coordinate Plane

EUREKA
MATH™

7. Figure $DEFG$ is shown on the coordinate plane below. The figure is dilated from the origin by scale factor $r = \frac{2}{3}$. Identify the coordinates of the dilated figure $D'E'F'G'$, and then draw and label figure $D'E'F'G'$ on the coordinate plane.

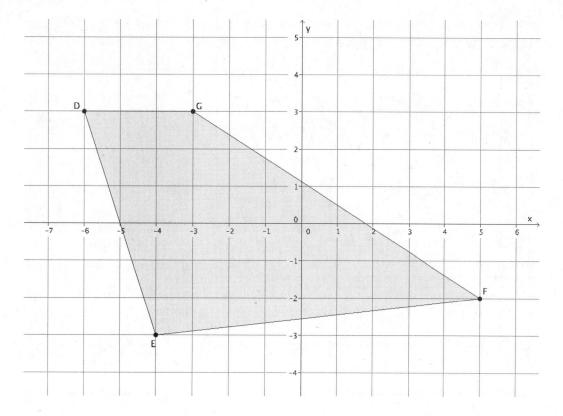

EUREKA
MATH™

8. The triangle ABC has coordinates $A(3,2)$, $B(12,3)$, and $C(9,12)$. Draw and label triangle ABC on the coordinate plane. The triangle is dilated from the origin by scale factor $r = \frac{1}{3}$. Identify the coordinates of the dilated triangle $A'B'C'$, and then draw and label triangle $A'B'C'$ on the coordinate plane.

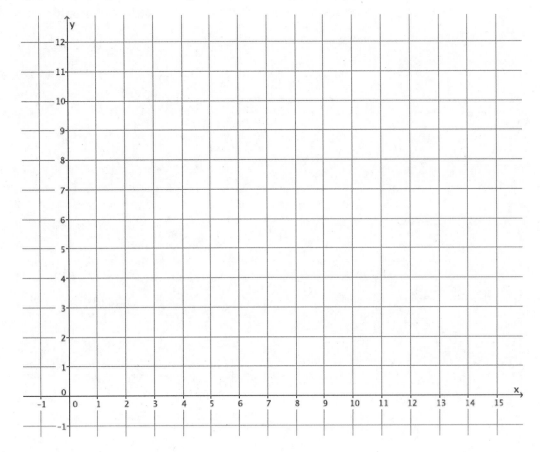

Lesson 6:   Dilations on the Coordinate Plane

EUREKA
MATH™

Lesson Summary

Dilation has a multiplicative effect on the coordinates of a point in the plane. Given a point (x, y) in the plane, a dilation from the origin with scale factor r moves the point (x, y) to (rx, ry).

For example, if a point $(3, -5)$ in the plane is dilated from the origin by a scale factor of $r = 4$, then the coordinates of the dilated point are $(4 \cdot 3, 4 \cdot (-5)) = (12, -20)$.

Problem Set

1. Triangle ABC is shown on the coordinate plane below. The triangle is dilated from the origin by scale factor $r = 4$. Identify the coordinates of the dilated triangle $A'B'C'$.

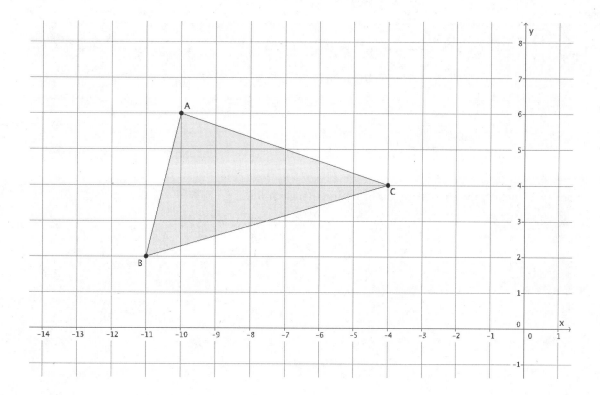

2. Triangle ABC is shown on the coordinate plane below. The triangle is dilated from the origin by scale factor $r = \frac{5}{4}$.
    Identify the coordinates of the dilated triangle $A'B'C'$.

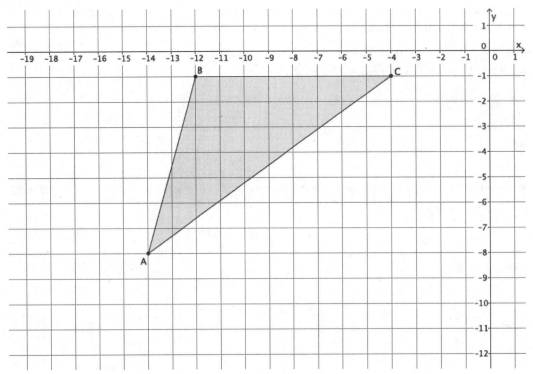

3. The triangle ABC has coordinates $A(6, 1)$, $B(12, 4)$, and $C(-6, 2)$. The triangle is dilated from the origin by a scale
    factor $r = \frac{1}{2}$. Identify the coordinates of the dilated triangle $A'B'C'$.

EUREKA
MATH™

4. Figure $DEFG$ is shown on the coordinate plane below. The figure is dilated from the origin by scale factor $r = \frac{3}{2}$. Identify the coordinates of the dilated figure $D'E'F'G'$, and then draw and label figure $D'E'F'G'$ on the coordinate plane.

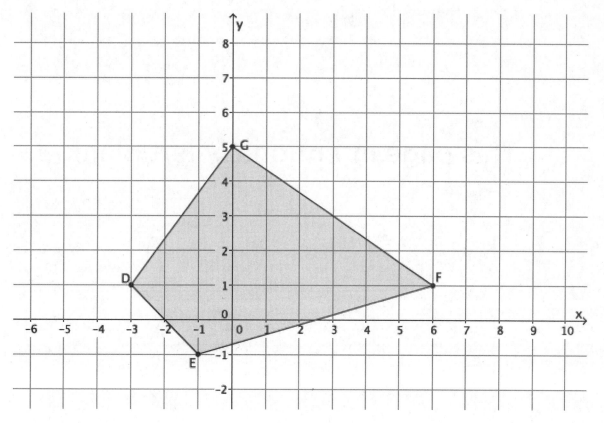

5. Figure $DEFG$ has coordinates $D(1, 1)$, $E(7, 3)$, $F(5, -4)$, and $G(-1, -4)$. The figure is dilated from the origin by scale factor $r = 7$. Identify the coordinates of the dilated figure $D'E'F'G'$.

This page intentionally left blank

Lesson 7: Informal Proofs of Properties of Dilation

Classwork

Exercise

Use the diagram below to prove the theorem: *Dilations preserve the measures of angles.*

Let there be a dilation from center O with scale factor r. Given $\angle PQR$, show that since $P' = Dilation(P)$, $Q' = Dilation(Q)$, and $R' = Dilation(R)$, then $|\angle PQR| = |\angle P'Q'R'|$. That is, show that the image of the angle after a dilation has the same measure, in degrees, as the original.

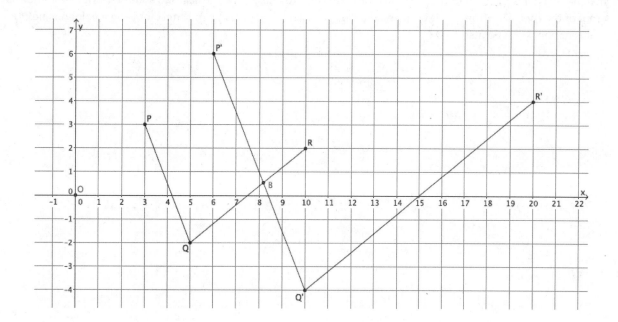

Problem Set

1. A dilation from center O by scale factor r of a line maps to what? Verify your claim on the coordinate plane.

2. A dilation from center O by scale factor r of a segment maps to what? Verify your claim on the coordinate plane.

3. A dilation from center O by scale factor r of a ray maps to what? Verify your claim on the coordinate plane.

4. Challenge Problem:

   Prove the theorem: *A dilation maps lines to lines.*

   Let there be a dilation from center O with scale factor r so that $P' = Dilation(P)$ and $Q' = Dilation(Q)$. Show that line PQ maps to line $P'Q'$ (i.e., that dilations map lines to lines). Draw a diagram, and then write your informal proof of the theorem. (Hint: This proof is a lot like the proof for segments. This time, let U be a point on line PQ that is not between points P and Q.)

Lesson 8: Similarity

Example 1

In the picture below, we have a triangle ABC that has been dilated from center O by a scale factor of $r = \frac{1}{2}$. It is noted by $A'B'C'$. We also have triangle $A''B''C''$, which is congruent to triangle $A'B'C'$ (i.e., $\triangle A'B'C' \cong \triangle A''B''C''$).

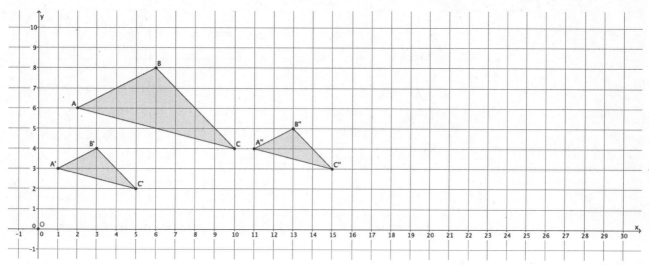

Describe the sequence that would map triangle $A''B''C''$ onto triangle ABC.

Exercises

1. Triangle ABC was dilated from center O by scale factor $r = \frac{1}{2}$. The dilated triangle is noted by $A'B'C'$. Another triangle $A''B''C''$ is congruent to triangle $A'B'C'$ (i.e., $\triangle A''B''C'' \cong \triangle A'B'C'$). Describe a dilation followed by the basic rigid motion that would map triangle $A''B''C''$ onto triangle ABC.

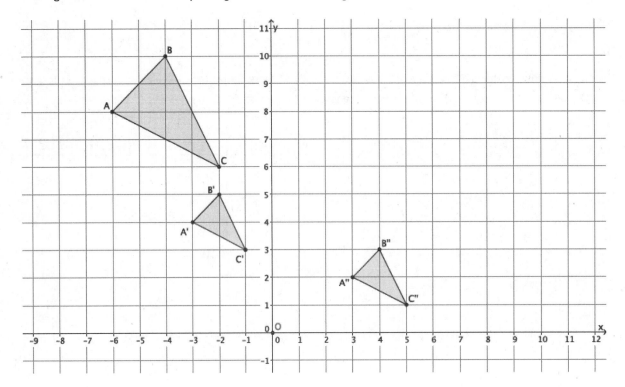

EUREKA
MATH™

2. Describe a sequence that would show △ $ABC \sim$ △ $A'B'C'$.

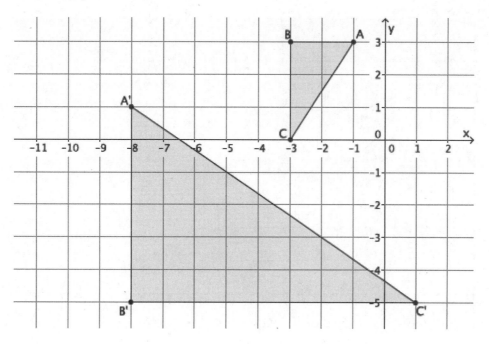

3. Are the two triangles shown below similar? If so, describe a sequence that would prove △ $ABC \sim$ △ $A'B'C'$. If not, state how you know they are not similar.

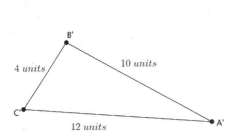

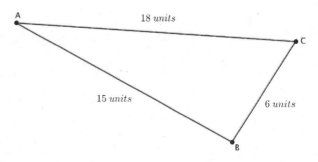

4. Are the two triangles shown below similar? If so, describe a sequence that would prove △ $ABC \sim$ △ $A'B'C'$. If not, state how you know they are not similar.

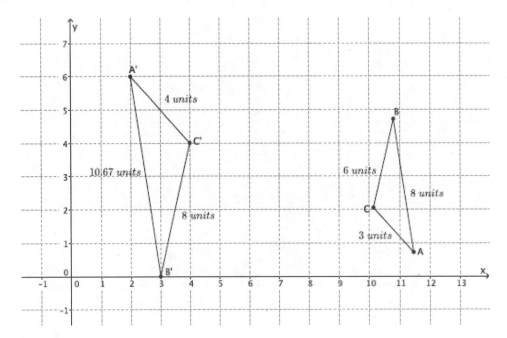

Lesson 8:   Similarity

EUREKA
MATH™

Lesson Summary

A *similarity transformation* (or a *similarity*) is a sequence of a finite number of dilations or basic rigid motions. Two figures are *similar* if there is a similarity transformation taking one figure onto the other figure. Every similarity can be represented as a dilation followed by a congruence.

The notation $\triangle ABC \sim \triangle A'B'C'$ means that $\triangle ABC$ is similar to $\triangle A'B'C'$.

Problem Set

1. In the picture below, we have triangle DEF that has been dilated from center O by scale factor $r = 4$. It is noted by $D'E'F'$. We also have triangle $D''E''F''$, which is congruent to triangle $D'E'F'$ (i.e., $\triangle D'E'F' \cong \triangle D''E''F''$). Describe the sequence of a dilation, followed by a congruence (of one or more rigid motions), that would map triangle $D''E''F''$ onto triangle DEF.

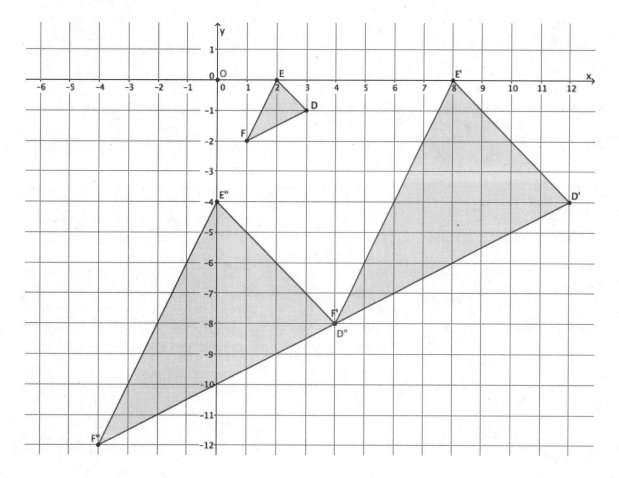

2. Triangle ABC was dilated from center O by scale factor $r = \frac{1}{2}$. The dilated triangle is noted by $A'B'C'$. Another triangle $A''B''C''$ is congruent to triangle $A'B'C'$ (i.e., $\triangle A''B''C'' \cong \triangle A'B'C'$). Describe the dilation followed by the basic rigid motions that would map triangle $A''B''C''$ onto triangle ABC.

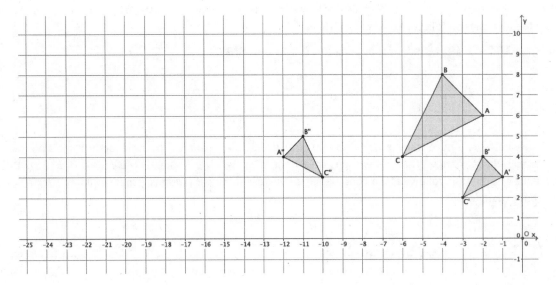

3. Are the two figures shown below similar? If so, describe a sequence that would prove the similarity. If not, state how you know they are not similar.

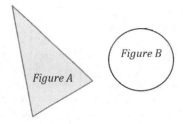

EUREKA MATH™

4. Triangle ABC is similar to triangle $A'B'C'$ (i.e., $\triangle ABC \sim \triangle A'B'C'$). Prove the similarity by describing a sequence that would map triangle $A'B'C'$ onto triangle ABC.

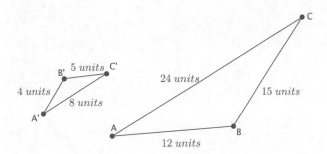

5. Are the two figures shown below similar? If so, describe a sequence that would prove $\triangle ABC \sim \triangle A'B'C'$. If not, state how you know they are not similar.

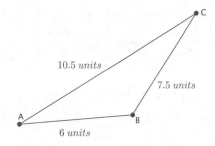

6.   Describe a sequence that would show $\triangle ABC \sim \triangle A'B'C'$.

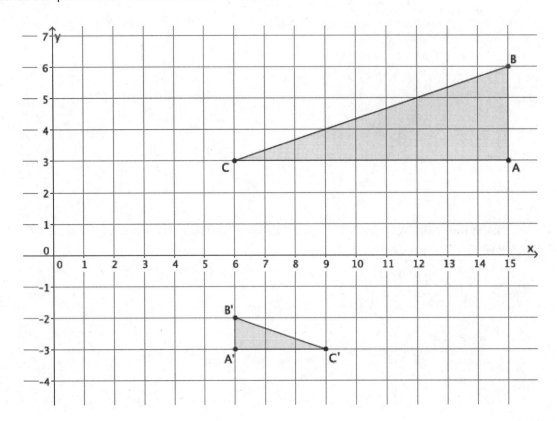

EUREKA
MATH™

Lesson 9: Basic Properties of Similarity

Exploratory Challenge 1

The goal is to show that if $\triangle ABC$ is similar to $\triangle A'B'C'$, then $\triangle A'B'C'$ is similar to $\triangle ABC$. Symbolically, if $\triangle ABC \sim \triangle A'B'C'$, then $\triangle A'B'C' \sim \triangle ABC$.

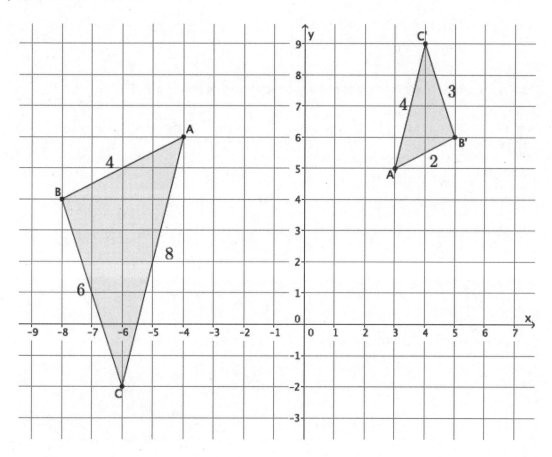

a.   First, determine whether or not $\triangle ABC$ is in fact similar to $\triangle A'B'C'$. (If it isn't, then no further work needs to be done.) Use a protractor to verify that the corresponding angles are congruent and that the ratios of the corresponding sides are equal to some scale factor.

b.   Describe the sequence of dilation followed by a congruence that proves $\triangle ABC \sim \triangle A'B'C'$.

c.   Describe the sequence of dilation followed by a congruence that proves $\triangle A'B'C' \sim \triangle ABC$.

d.   Is it true that $\triangle ABC \sim \triangle A'B'C'$ and $\triangle A'B'C' \sim \triangle ABC$? Why do you think this is so?

EUREKA
MATH

Exploratory Challenge 2

The goal is to show that if $\triangle ABC$ is similar to $\triangle A'B'C'$ and $\triangle A'B'C'$ is similar to $\triangle A''B''C''$, then $\triangle ABC$ is similar to $\triangle A''B''C''$. Symbolically, if $\triangle ABC \sim \triangle A'B'C'$ and $\triangle A'B'C' \sim \triangle A''B''C''$, then $\triangle ABC \sim \triangle A''B''C''$.

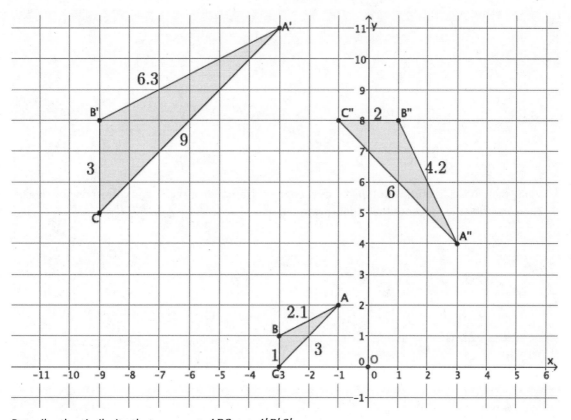

a. Describe the similarity that proves $\triangle ABC \sim \triangle A'B'C'$.

b. Describe the similarity that proves $\triangle A'B'C' \sim \triangle A''B''C''$.

c. Verify that, in fact, $\triangle ABC \sim \triangle A''B''C''$ by checking corresponding angles and corresponding side lengths. Then, describe the sequence that would prove the similarity $\triangle ABC \sim \triangle A''B''C''$.

d. Is it true that if $\triangle ABC \sim \triangle A'B'C'$ and $\triangle A'B'C' \sim \triangle A''B''C''$, then $\triangle ABC \sim \triangle A''B''C''$? Why do you think this is so?

EUREKA
MATH™

Problem Set

1. Would a dilation alone be enough to show that similarity is symmetric? That is, would a dilation alone prove that if $\triangle ABC \sim \triangle A'B'C'$, then $\triangle A'B'C' \sim \triangle ABC$? Consider the two examples below.

    a. Given $\triangle ABC \sim \triangle A'B'C'$, is a dilation enough to show that $\triangle A'B'C' \sim \triangle ABC$? Explain.

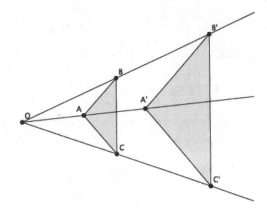

    b. Given $\triangle ABC \sim \triangle A'B'C'$, is a dilation enough to show that $\triangle A'B'C' \sim \triangle ABC$? Explain.

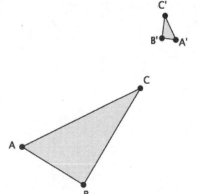

    c. In general, is dilation enough to prove that similarity is a symmetric relation? Explain.

2. Would a dilation alone be enough to show that similarity is transitive? That is, would a dilation alone prove that if
    $\triangle ABC \sim \triangle A'B'C'$ and $\triangle A'B'C' \sim \triangle A''B''C''$, then $\triangle ABC \sim \triangle A''B''C''$? Consider the two examples below.

    a. Given $\triangle ABC \sim \triangle A'B'C'$ and $\triangle A'B'C' \sim \triangle A''B''C''$, is a dilation enough to show that $\triangle ABC \sim \triangle A''B''C''$?
        Explain.

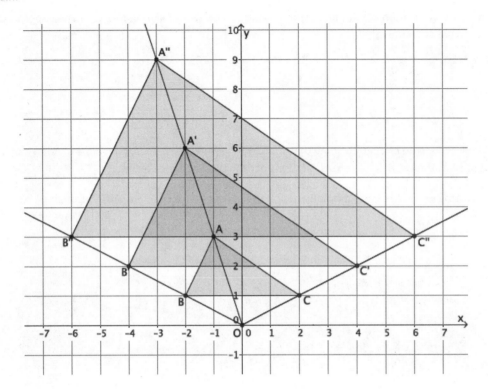

Lesson 9:   Basic Properties of Similarity

**EUREKA
MATH™**

b.   Given △ $ABC \sim$ △ $A'B'C'$ and △ $A'B'C' \sim$ △ $A''B''C''$, is a dilation enough to show that △ $ABC \sim$ △ $A''B''C''$? Explain.

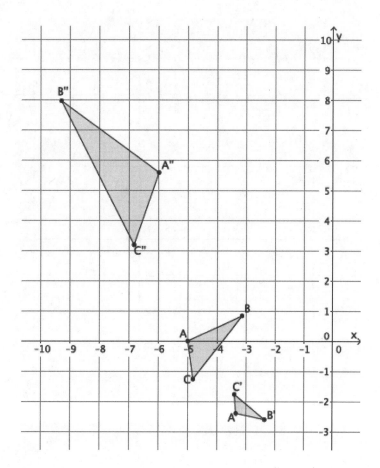

c.   In general, is dilation enough to prove that similarity is a transitive relation? Explain.

EUREKA MATH™

© 2015 Great Minds. eureka-math.org
G8-M3M4M5-SE-B2-1.3.1-01.2016

3. In the diagram below, △ ABC ~△ $A'B'C'$ and △ $A'B'C'$ ~ △ $A''B''C''$. Is △ ABC ~ △ $A''B''C''$? If so, describe the dilation followed by the congruence that demonstrates the similarity.

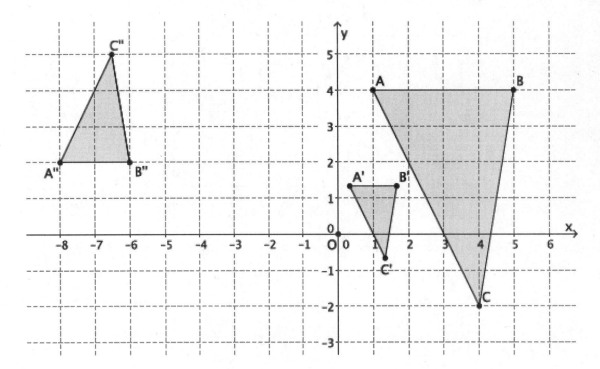

EUREKA
MATH™

Lesson 10: Informal Proof of AA Criterion for Similarity

Classwork

Exercises 1–5

1. Use a protractor to draw a pair of triangles with two pairs of interior angles that are equal in measure. Then, measure the lengths of the sides, and verify that the lengths of their corresponding sides are equal in ratio.

2. Draw a new pair of triangles with two pairs of interior angles that are equal in measure. Then, measure the lengths of the sides, and verify that the lengths of their corresponding sides are equal in ratio.

3. Are the triangles shown below similar? Present an informal argument as to why they are or are not similar.

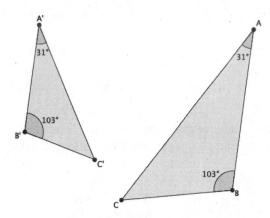

4. Are the triangles shown below similar? Present an informal argument as to why they are or are not similar.

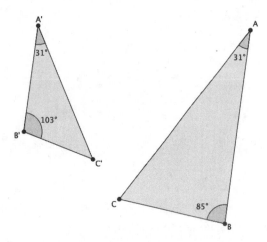

5. Are the triangles shown below similar? Present an informal argument as to why they are or are not similar.

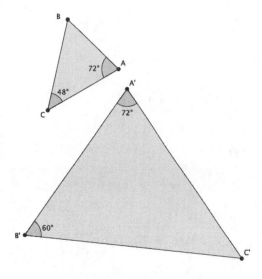

Lesson 10:   Informal Proof of AA Criterion for Similarity

EUREKA
MATH

Problem Set

1. Are the triangles shown below similar? Present an informal argument as to why they are or are not similar.

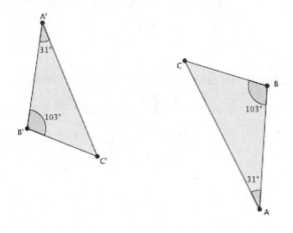

2. Are the triangles shown below similar? Present an informal argument as to why they are or are not similar.

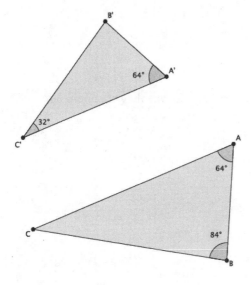

3.   Are the triangles shown below similar? Present an informal argument as to why they are or are not similar.

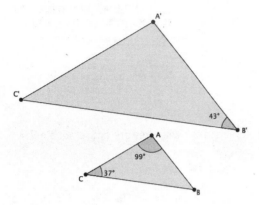

4.   Are the triangles shown below similar? Present an informal argument as to why they are or are not similar.

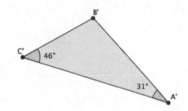

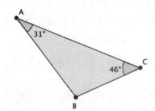

5.   Are the triangles shown below similar? Present an informal argument as to why they are or are not similar.

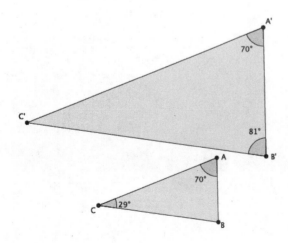

EUREKA
MATH™

6. Are the triangles shown below similar? Present an informal argument as to why they are or are not similar.

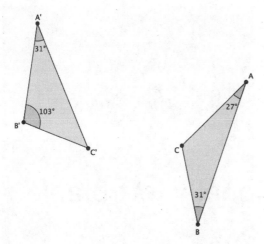

7. Are the triangles shown below similar? Present an informal argument as to why they are or are not similar.

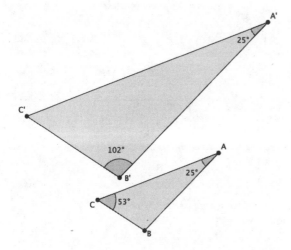

EUREKA
MATH™

Lesson 10:   Informal Proof of AA Criterion for Similarity

S.57

© 2015 Great Minds. eureka-math.org
G8-M3M4M5-SE-B2-1.3.1-01.2016

This page intentionally left blank

Lesson 11: More About Similar Triangles

Exercises

1. In the diagram below, you have $\triangle ABC$ and $\triangle AB'C'$. Use this information to answer parts (a)–(d).

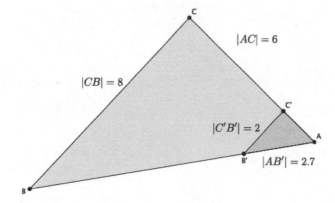

$|AC| = 6$

$|CB| = 8$

$|C'B'| = 2$

$|AB'| = 2.7$

   a. Based on the information given, is $\triangle ABC \sim \triangle AB'C'$? Explain.

   b. Assume the line containing BC is parallel to the line containing $B'C'$. With this information, can you say that $\triangle ABC \sim \triangle AB'C'$? Explain.

   c. Given that $\triangle ABC \sim \triangle AB'C'$, determine the length of side $\overline{AC'}$.

   d. Given that $\triangle ABC \sim \triangle AB'C'$, determine the length of side $\overline{AB}$.

EUREKA
MATH™

Lesson 11: More About Similar Triangles

S.59

© 2015 Great Minds. eureka-math.org
G8-M3M4M5-SE-B2-1.3.1-01.2016

2. In the diagram below, you have △ ABC and △ $A'B'C'$. Use this information to answer parts (a)–(c).

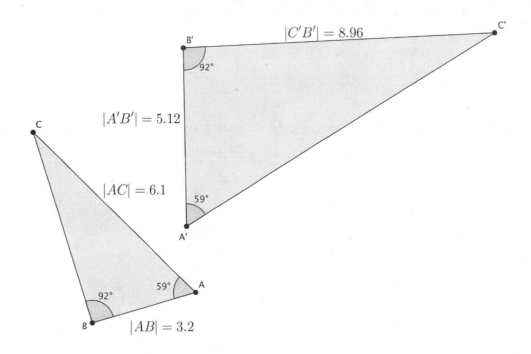

a. Based on the information given, is △ $ABC \sim$ △ $A'B'C'$? Explain.

b. Given that △ $ABC \sim$ △ $A'B'C'$, determine the length of side $\overline{A'C'}$.

c. Given that △ $ABC \sim$ △ $A'B'C'$, determine the length of side $\overline{BC}$.

EUREKA
MATH™

3.   In the diagram below, you have △ ABC and △ $A'B'C'$. Use this information to answer the question below.

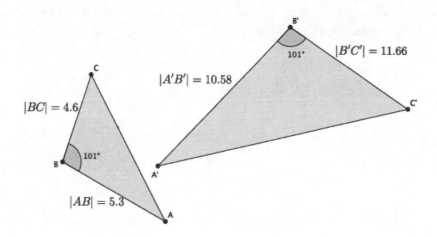

Based on the information given, is △ ABC ~ △ $A'B'C'$? Explain.

EUREKA
MATH™

Lesson 11:   More About Similar Triangles

S.61

© 2015 Great Minds. eureka-math.org
G8-M3M4M5-SE-B2-1.3.1-01.2016

Lesson Summary

Given just one pair of corresponding angles of a triangle as equal in measure, use the side lengths along the given angle to determine if the triangles are in fact similar.

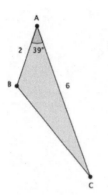 

$|\angle A| = |\angle D|$ and $\dfrac{1}{2} = \dfrac{3}{6} = r$; therefore, $\triangle ABC \sim \triangle DEF$.

Given similar triangles, use the fact that ratios of corresponding sides are equal to find any missing measurements.

Problem Set

1. In the diagram below, you have $\triangle ABC$ and $\triangle A'B'C'$. Use this information to answer parts (a)–(b).

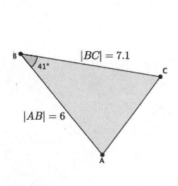 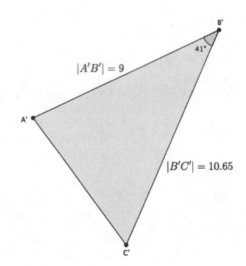

a. Based on the information given, is $\triangle ABC \sim \triangle A'B'C'$? Explain.

b. Assume the length of side $\overline{AC}$ is 4.3. What is the length of side $\overline{A'C'}$?

Lesson 11:   More About Similar Triangles

EUREKA
MATH™

2. In the diagram below, you have $\triangle ABC$ and $\triangle AB'C'$. Use this information to answer parts (a)–(d).

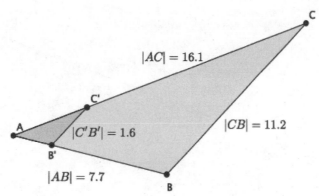

a. Based on the information given, is $\triangle ABC \sim \triangle AB'C'$? Explain.

b. Assume the line containing $\overline{BC}$ is parallel to the line containing $\overline{B'C'}$. With this information, can you say that $\triangle ABC \sim \triangle AB'C'$? Explain.

c. Given that $\triangle ABC \sim \triangle AB'C'$, determine the length of side $\overline{AC'}$.

d. Given that $\triangle ABC \sim \triangle AB'C'$, determine the length of side $\overline{AB'}$.

3. In the diagram below, you have $\triangle ABC$ and $\triangle A'B'C'$. Use this information to answer parts (a)–(c).

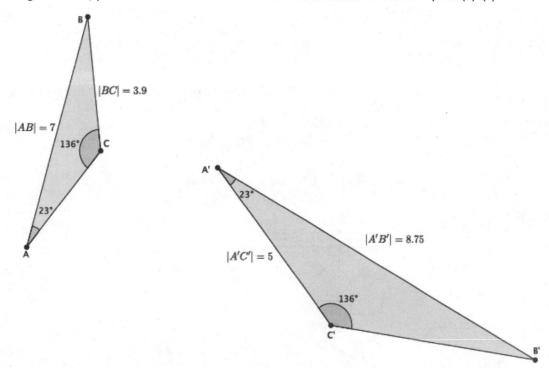

a. Based on the information given, is $\triangle ABC \sim \triangle A'B'C'$? Explain.

b. Given that $\triangle ABC \sim \triangle A'B'C'$, determine the length of side $\overline{B'C'}$.

c. Given that $\triangle ABC \sim \triangle A'B'C'$, determine the length of side $\overline{AC}$.

4.   In the diagram below, you have △ ABC and △ $AB'C'$. Use this information to answer the question below.

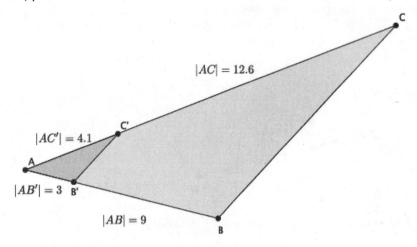

Based on the information given, is △ ABC ~ △ $AB'C'$? Explain.

5.   In the diagram below, you have △ ABC and △ $A'B'C'$. Use this information to answer parts (a)–(b).

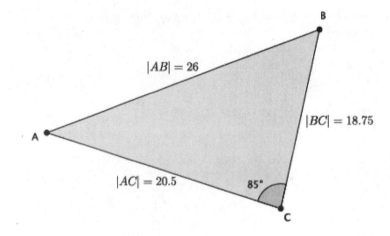

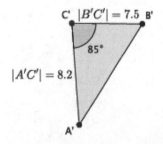

a.   Based on the information given, is △ ABC ~ △ $A'B'C'$? Explain.

b.   Given that △ ABC ~ △ $A'B'C'$, determine the length of side $\overline{A'B'}$.

EUREKA
MATH™

Lesson 12: Modeling Using Similarity

Example

Not all flagpoles are perfectly *upright* (i.e., perpendicular to the ground). Some are oblique (i.e., neither parallel nor at a right angle, slanted). Imagine an oblique flagpole in front of an abandoned building. The question is, can we use sunlight and shadows to determine the length of the flagpole?

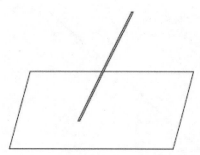

Assume that we know the following information: The length of the shadow of the flagpole is 15 feet. There is a mark on the flagpole 3 feet from its base. The length of the shadow of this three-foot portion of the flagpole is 1.7 feet.

Mathematical Modeling Exercises 1–3

1. You want to determine the approximate height of one of the tallest buildings in the city. You are told that if you place a mirror some distance from yourself so that you can see the top of the building in the mirror, then you can indirectly measure the height using similar triangles. Let point O be the location of the mirror so that the person shown can see the top of the building.

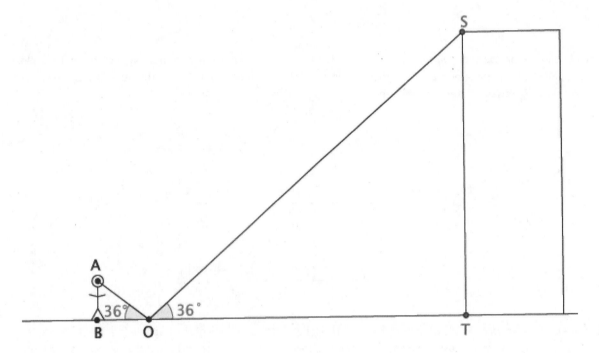

a. Explain why $\triangle ABO \sim \triangle STO$.

b. Label the diagram with the following information: The distance from eye level straight down to the ground is 5.3 feet. The distance from the person to the mirror is 7.2 feet. The distance from the person to the base of the building is 1,750 feet. The height of the building is represented by x.

c. What is the distance from the mirror to the building?

Lesson 12:   Modeling Using Similarity

EUREKA
MATH™

d. Do you have enough information to determine the approximate height of the building? If yes, determine the approximate height of the building. If not, what additional information is needed?

2. A geologist wants to determine the distance across the widest part of a nearby lake. The geologist marked off specific points around the lake so that the line containing $\overline{DE}$ would be parallel to the line containing $\overline{BC}$. The segment BC is selected specifically because it is the widest part of the lake. The segment DE is selected specifically because it is a short enough distance to easily measure. The geologist sketched the situation as shown below.

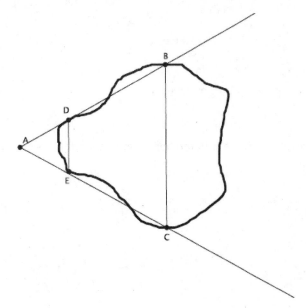

a. Has the geologist done enough work so far to use similar triangles to help measure the widest part of the lake? Explain.

EUREKA
MATH™

Lesson 12:   Modeling Using Similarity

S.67

© 2015 Great Minds. eureka-math.org
G8-M3M4M5-SE-B2-1.3.1-01.2016

b.   The geologist has made the following measurements: $|DE| = 5$ feet, $|AE| = 7$ feet, and $|EC| = 15$ feet. Does she have enough information to complete the task? If so, determine the length across the widest part of the lake. If not, state what additional information is needed.

c.   Assume the geologist could only measure a maximum distance of 12 feet. Could she still find the distance across the widest part of the lake? What would need to be done differently?

3.   A tree is planted in the backyard of a house with the hope that one day it is tall enough to provide shade to cool the house. A sketch of the house, tree, and sun is shown below.

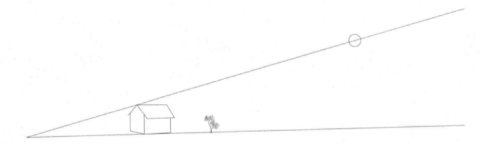

a.   What information is needed to determine how tall the tree must be to provide the desired shade?

EUREKA
MATH™

b. Assume that the sun casts a shadow 32 feet long from a point on top of the house to a point in front of the house. The distance from the end of the house's shadow to the base of the tree is 53 feet. If the house is 16 feet tall, how tall must the tree get to provide shade for the house?

c. Assume that the tree grows at a rate of 2.5 feet per year. If the tree is now 7 feet tall, about how many years will it take for the tree to reach the desired height?

Problem Set

1. The world's tallest living tree is a redwood in California. It's about 370 feet tall. In a local park, there is a very tall tree. You want to find out if the tree in the local park is anywhere near the height of the famous redwood.

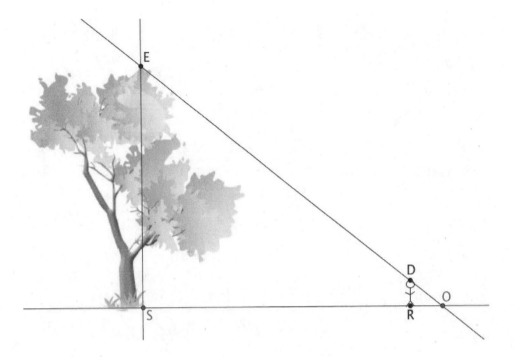

a. Describe the triangles in the diagram, and explain how you know they are similar or not.

b. Assume $\triangle ESO \sim \triangle DRO$. A friend stands in the shadow of the tree. He is exactly 5.5 feet tall and casts a shadow of 12 feet. Is there enough information to determine the height of the tree? If so, determine the height. If not, state what additional information is needed.

c. Your friend stands exactly 477 feet from the base of the tree. Given this new information, determine about how many feet taller the world's tallest tree is compared to the one in the local park.

d. Assume that your friend stands in the shadow of the world's tallest redwood, and the length of his shadow is just 8 feet long. How long is the shadow cast by the tree?

2. A reasonable skateboard ramp makes a 25° angle with the ground. A two-foot-tall ramp requires about 4.3 feet of wood along the base and about 4.7 feet of wood from the ground to the top of the two-foot height to make the ramp.

a. Sketch a diagram to represent the situation.

b. Your friend is a daredevil and has decided to build a ramp that is 5 feet tall. What length of wood is needed to make the base of the ramp? Explain your answer using properties of similar triangles.

c. What length of wood is required to go from the ground to the top of the 5-foot height to make the ramp? Explain your answer using properties of similar triangles.

Lesson 12: Modeling Using Similarity

Lesson 13: Proof of the Pythagorean Theorem

Classwork

Exercises

Use the Pythagorean theorem to determine the unknown length of the right triangle.

1. Determine the length of side c in each of the triangles below.

    a.

    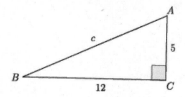

    b.

    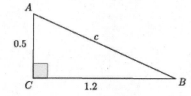

2. Determine the length of side b in each of the triangles below.

    a.

    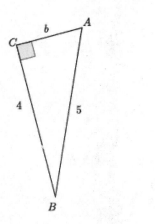

    b.

    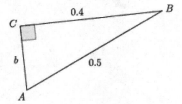

3.   Determine the length of $\overline{QS}$. (Hint: Use the Pythagorean theorem twice.)

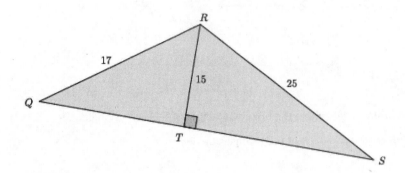

EUREKA
MATH

Problem Set

Use the Pythagorean theorem to determine the unknown length of the right triangle.

1. Determine the length of side c in each of the triangles below.

    a.

    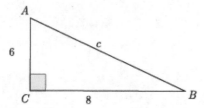

    b.

    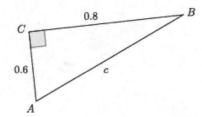

2. Determine the length of side a in each of the triangles below.

    a.

    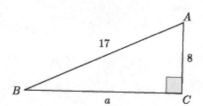

    b.

    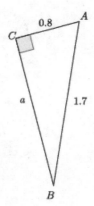

3.   Determine the length of side b in each of the triangles below.

a.

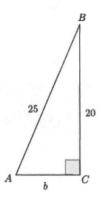

b.

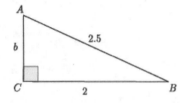

4.   Determine the length of side a in each of the triangles below.

a.

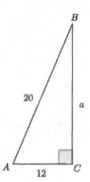

b.

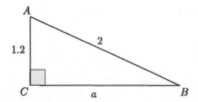

5.   What did you notice in each of the pairs of Problems 1–4? How might what you noticed be helpful in solving problems like these?

EUREKA
MATH

Lesson 14: The Converse of the Pythagorean Theorem

Classwork

Exercises

1. The numbers in the diagram below indicate the units of length of each side of the triangle. Is the triangle shown below a right triangle? Show your work, and answer in a complete sentence.

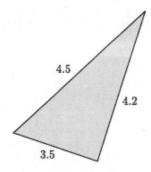

2. The numbers in the diagram below indicate the units of length of each side of the triangle. Is the triangle shown below a right triangle? Show your work, and answer in a complete sentence.

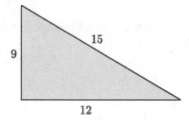

3. The numbers in the diagram below indicate the units of length of each side of the triangle. Is the triangle shown below a right triangle? Show your work, and answer in a complete sentence.

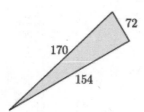

© 2015 Great Minds. eureka-math.org
G8-M3M4M5-SE-B2-1.3.1-01.2016

4. The numbers in the diagram below indicate the units of length of each side of the triangle. Is the triangle shown below a right triangle? Show your work, and answer in a complete sentence.

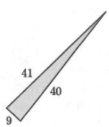

5. The numbers in the diagram below indicate the units of length of each side of the triangle. Is the triangle shown below a right triangle? Show your work, and answer in a complete sentence.

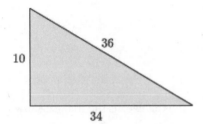

6. The numbers in the diagram below indicate the units of length of each side of the triangle. Is the triangle shown below a right triangle? Show your work, and answer in a complete sentence.

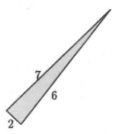

7. The numbers in the diagram below indicate the units of length of each side of the triangle. Is the triangle shown below a right triangle? Show your work, and answer in a complete sentence.

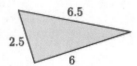

Lesson Summary

The converse of the Pythagorean theorem states that if the side lengths of a triangle, a, b, c, satisfy $a^2 + b^2 = c^2$, then the triangle is a right triangle.

If the side lengths of a triangle, a, b, c, do not satisfy $a^2 + b^2 = c^2$, then the triangle is not a right triangle.

Problem Set

1. The numbers in the diagram below indicate the units of length of each side of the triangle. Is the triangle shown below a right triangle? Show your work, and answer in a complete sentence.

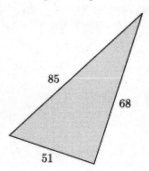

2. The numbers in the diagram below indicate the units of length of each side of the triangle. Is the triangle shown below a right triangle? Show your work, and answer in a complete sentence.

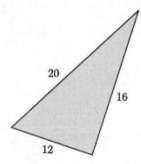

3. The numbers in the diagram below indicate the units of length of each side of the triangle. Is the triangle shown below a right triangle? Show your work, and answer in a complete sentence.

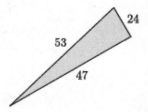

4. The numbers in the diagram below indicate the units of length of each side of the triangle. Sam said that the following triangle is a right triangle because $9 + 32 = 40$. Explain to Sam what he did wrong to reach this conclusion and what the correct solution is.

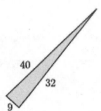

5. The numbers in the diagram below indicate the units of length of each side of the triangle. Is the triangle shown below a right triangle? Show your work, and answer in a complete sentence.

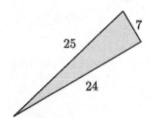

6. Jocelyn said that the triangle below is not a right triangle. Her work is shown below. Explain what she did wrong, and show Jocelyn the correct solution.

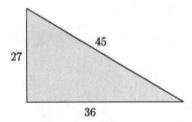

We need to check if $27^2 + 45^2 = 36^2$ is a true statement. The left side of the equation is equal to 2,754. The right side of the equation is equal to 1,296. That means $27^2 + 45^2 = 36^2$ is not true, and the triangle shown is not a right triangle.

EUREKA MATH™

Eureka Math
Grade 8
Module 4

Special thanks go to the Gordon A. Cain Center and to the Department of Mathematics at Louisiana State University for their support in the development of *Eureka Math*.

For a free *Eureka Math* Teacher Resource Pack, Parent Tip Sheets, and more please visit www.Eureka.tools

ISBN 978-1-63255-321-8

Lesson 1: Writing Equations Using Symbols

Classwork

Exercises

Write each of the following statements using symbolic language.

1. The sum of four consecutive even integers is −28.

2. A number is four times larger than the square of half the number.

3. Steven has some money. If he spends $9.00, then he will have $\frac{3}{5}$ of the amount he started with.

4. The sum of a number squared and three less than twice the number is 129.

5. Miriam read a book with an unknown number of pages. The first week, she read five less than $\frac{1}{3}$ of the pages. The second week, she read 171 more pages and finished the book. Write an equation that represents the total number of pages in the book.

Lesson Summary

Begin all word problems by defining your variables. State clearly what you want each symbol to represent.

Written mathematical statements can be represented as more than one correct symbolic statement.

Break complicated problems into smaller parts, or try working them with simpler numbers.

Problem Set

Write each of the following statements using symbolic language.

1. Bruce bought two books. One book costs $4.00 more than three times the other. Together, the two books cost him $72.

2. Janet is three years older than her sister Julie. Janet's brother is eight years younger than their sister Julie. The sum of all of their ages is 55 years.

3. The sum of three consecutive integers is 1,623.

4. One number is six more than another number. The sum of their squares is 90.

5. When you add 18 to $\frac{1}{4}$ of a number, you get the number itself.

6. When a fraction of 17 is taken away from 17, what remains exceeds one-third of seventeen by six.

7. The sum of two consecutive even integers divided by four is 189.5.

8. Subtract seven more than twice a number from the square of one-third of the number to get zero.

9. The sum of three consecutive integers is 42. Let x be the middle of the three integers. Transcribe the statement accordingly.

© 2015 Great Minds. eureka-math.org
G8-M3M4M5-SE-B2-1.3.1-01.2016

Lesson 2: Linear and Nonlinear Expressions in x

Classwork

Exercises

Write each of the following statements in Exercises 1–12 as a mathematical expression. State whether or not the expression is linear or nonlinear. If it is nonlinear, then explain why.

1. The sum of a number and four times the number

2. The product of five and a number

3. Multiply six and the reciprocal of the quotient of a number and seven.

4. Twice a number subtracted from four times a number, added to 15

5. The square of the sum of six and a number

6. The cube of a positive number divided by the square of the same positive number

7. The sum of four consecutive numbers

8. Four subtracted from the reciprocal of a number

9. Half of the product of a number multiplied by itself three times

10. The sum that shows how many pages Maria read if she read 45 pages of a book yesterday and $\frac{2}{3}$ of the remaining pages today

11. An admission fee of $10 plus an additional $2 per game

12. Five more than four times a number and then twice that sum

EUREKA
MATH™

> **Lesson Summary**
>
> A *linear expression* is an expression that is equivalent to the sum or difference of one or more expressions where each expression is either a number, a variable, or a product of a number and a variable.
>
> A linear expression in x can be represented by terms whose variable x is raised to either a power of 0 or 1. For example, $4 + 3x$, $7x + x - 15$, and $\frac{1}{2}x + 7 - 2$ are all linear expressions in x. A nonlinear expression in x has terms where x is raised to a power that is not 0 or 1. For example, $2x^2 - 9$, $-6x^{-3} + 8 + x$, and $\frac{1}{x} + 8$ are all nonlinear expressions in x.

Problem Set

Write each of the following statements as a mathematical expression. State whether the expression is linear or nonlinear. If it is nonlinear, then explain why.

1. A number decreased by three squared

2. The quotient of two and a number, subtracted from seventeen

3. The sum of thirteen and twice a number

4. 5.2 more than the product of seven and a number

5. The sum that represents the number of tickets sold if 35 tickets were sold Monday, half of the remaining tickets were sold on Tuesday, and 14 tickets were sold on Wednesday

6. The product of 19 and a number, subtracted from the reciprocal of the number cubed

7. The product of 15 and a number, and then the product multiplied by itself four times

8. A number increased by five and then divided by two

9. Eight times the result of subtracting three from a number

10. The sum of twice a number and four times a number subtracted from the number squared

11. One-third of the result of three times a number that is increased by 12

12. Five times the sum of one-half and a number

13. Three-fourths of a number multiplied by seven

14. The sum of a number and negative three, multiplied by the number

15. The square of the difference between a number and 10

EUREKA
MATH™

Lesson 3: Linear Equations in x

Classwork

Exercises

1. Is the equation a true statement when $x = -3$? In other words, is -3 a solution to the equation $6x + 5 = 5x + 8 + 2x$? Explain.

2. Does $x = 12$ satisfy the equation $16 - \frac{1}{2}x = \frac{3}{4}x + 1$? Explain.

3. Chad solved the equation $24x + 4 + 2x = 3(10x - 1)$ and is claiming that $x = 2$ makes the equation true. Is Chad correct? Explain.

4. Lisa solved the equation $x + 6 = 8 + 7x$ and claimed that the solution is $x = -\frac{1}{3}$. Is she correct? Explain.

5. Angel transformed the following equation from $6x + 4 - x = 2(x + 1)$ to $10 = 2(x + 1)$. He then stated that the solution to the equation is $x = 4$. Is he correct? Explain.

6. Claire was able to verify that $x = 3$ was a solution to her teacher's linear equation, but the equation got erased from the board. What might the equation have been? Identify as many equations as you can with a solution of $x = 3$.

7. Does an equation always have a solution? Could you come up with an equation that does not have a solution?

EUREKA
MATH™

Lesson Summary

An equation is a statement about equality between two expressions. If the expression on the left side of the equal sign has the same value as the expression on the right side of the equal sign, then you have a true equation.

A solution of a linear equation in x is a number, such that when all instances of x are replaced with the number, the left side will equal the right side. For example, 2 is a solution to $3x + 4 = x + 8$ because when $x = 2$, the left side of the equation is

$$3x + 4 = 3(2) + 4$$
$$= 6 + 4$$
$$= 10,$$

and the right side of the equation is

$$x + 8 = 2 + 8$$
$$= 10.$$

Since $10 = 10$, then $x = 2$ is a solution to the linear equation $3x + 4 = x + 8$.

Problem Set

1. Given that $2x + 7 = 27$ and $3x + 1 = 28$, does $2x + 7 = 3x + 1$? Explain.

2. Is -5 a solution to the equation $6x + 5 = 5x + 8 + 2x$? Explain.

3. Does $x = 1.6$ satisfy the equation $6 - 4x = -\dfrac{x}{4}$? Explain.

4. Use the linear equation $3(x + 1) = 3x + 3$ to answer parts (a)–(d).

   a. Does $x = 5$ satisfy the equation above? Explain.

   b. Is $x = -8$ a solution of the equation above? Explain.

   c. Is $x = \dfrac{1}{2}$ a solution of the equation above? Explain.

   d. What interesting fact about the equation $3(x + 1) = 3x + 3$ is illuminated by the answers to parts (a), (b), and (c)? Why do you think this is true?

This page intentionally left blank

Lesson 4: Solving a Linear Equation

Classwork

Exercises

For each problem, show your work, and check that your solution is correct.

1. Solve the linear equation $x + x + 2 + x + 4 + x + 6 = -28$. State the property that justifies your first step and why you chose it.

2. Solve the linear equation $2(3x + 2) = 2x - 1 + x$. State the property that justifies your first step and why you chose it.

3. Solve the linear equation $x - 9 = \frac{3}{5}x$. State the property that justifies your first step and why you chose it.

4. Solve the linear equation $29 - 3x = 5x + 5$. State the property that justifies your first step and why you chose it.

5. Solve the linear equation $\frac{1}{3}x - 5 + 171 = x$. State the property that justifies your first step and why you chose it.

EUREKA
MATH™

Lesson Summary

The properties of equality, shown below, are used to transform equations into simpler forms. If A, B, C are rational numbers, then:

- If $A = B$, then $A + C = B + C$.   Addition property of equality
- If $A = B$, then $A - C = B - C$.   Subtraction property of equality
- If $A = B$, then $A \cdot C = B \cdot C$.   Multiplication property of equality
- If $A = B$, then $\dfrac{A}{C} = \dfrac{B}{C}$, where C is not equal to zero. Division property of equality

To solve an equation, transform the equation until you get to the form of x equal to a constant ($x = 5$, for example).

Problem Set

For each problem, show your work, and check that your solution is correct.

1. Solve the linear equation $x + 4 + 3x = 72$. State the property that justifies your first step and why you chose it.

2. Solve the linear equation $x + 3 + x - 8 + x = 55$. State the property that justifies your first step and why you chose it.

3. Solve the linear equation $\dfrac{1}{2}x + 10 = \dfrac{1}{4}x + 54$. State the property that justifies your first step and why you chose it.

4. Solve the linear equation $\dfrac{1}{4}x + 18 = x$. State the property that justifies your first step and why you chose it.

5. Solve the linear equation $17 - x = \dfrac{1}{3} \cdot 15 + 6$. State the property that justifies your first step and why you chose it.

6. Solve the linear equation $\dfrac{x+x+2}{4} = 189.5$. State the property that justifies your first step and why you chose it.

7. Alysha solved the linear equation $2x - 3 - 8x = 14 + 2x - 1$. Her work is shown below. When she checked her answer, the left side of the equation did not equal the right side. Find and explain Alysha's error, and then solve the equation correctly.

$$2x - 3 - 8x = 14 + 2x - 1$$
$$-6x - 3 = 13 + 2x$$
$$-6x - 3 + 3 = 13 + 3 + 2x$$
$$-6x = 16 + 2x$$
$$-6x + 2x = 16$$
$$-4x = 16$$
$$\frac{-4}{-4}x = \frac{16}{-4}$$
$$x = -4$$

Lesson 4:   Solving a Linear Equation

EUREKA
MATH™

Lesson 5: Writing and Solving Linear Equations

Classwork

Example 1

One angle is five degrees less than three times the degree measure of another angle. Together, the angle measures have a sum of 143°. What is the measure of each angle?

Example 2

Given a right triangle, find the degree measure of the angles if one angle is ten degrees more than four times the degree measure of the other angle and the third angle is the right angle.

Exercises

For each of the following problems, write an equation and solve.

1. A pair of congruent angles are described as follows: The degree measure of one angle is three more than twice a number, and the other angle's degree measure is 54.5 less than three times the number. Determine the measure of the angles in degrees.

2. The measure of one angle is described as twelve more than four times a number. Its supplement is twice as large. Find the measure of each angle in degrees.

3. A triangle has angles described as follows: The measure of the first angle is four more than seven times a number, the measure of the second angle is four less than the first, and the measure of the third angle is twice as large as the first. What is the measure of each angle in degrees?

EUREKA
MATH™

4. One angle measures nine more than six times a number. A sequence of rigid motions maps the angle onto another angle that is described as being thirty less than nine times the number. What is the measure of the angle in degrees?

5. A right triangle is described as having an angle of measure six less than negative two times a number, another angle measure that is three less than negative one-fourth the number, and a right angle. What are the measures of the angles in degrees?

6. One angle is one less than six times the measure of another. The two angles are complementary angles. Find the measure of each angle in degrees.

Problem Set

For each of the following problems, write an equation and solve.

1. The measure of one angle is thirteen less than five times the measure of another angle. The sum of the measures of the two angles is 140°. Determine the measure of each angle in degrees.

2. An angle measures seventeen more than three times a number. Its supplement is three more than seven times the number. What is the measure of each angle in degrees?

3. The angles of a triangle are described as follows: $\angle A$ is the largest angle; its measure is twice the measure of $\angle B$. The measure of $\angle C$ is 2 less than half the measure of $\angle B$. Find the measures of the three angles in degrees.

4. A pair of corresponding angles are described as follows: The measure of one angle is five less than seven times a number, and the measure of the other angle is eight more than seven times the number. Are the angles congruent? Why or why not?

5. The measure of one angle is eleven more than four times a number. Another angle is twice the first angle's measure. The sum of the measures of the angles is 195°. What is the measure of each angle in degrees?

6. Three angles are described as follows: $\angle B$ is half the size of $\angle A$. The measure of $\angle C$ is equal to one less than two times the measure of $\angle B$. The sum of $\angle A$ and $\angle B$ is 114°. Can the three angles form a triangle? Why or why not?

Lesson 6: Solutions of a Linear Equation

Exercises

Find the value of x that makes the equation true.

1. $17 - 5(2x - 9) = -(-6x + 10) + 4$

2. $-(x - 7) + \frac{5}{3} = 2(x + 9)$

3. $\frac{4}{9} + 4(x - 1) = \frac{28}{9} - (x - 7x) + 1$

4. $5(3x + 4) - 2x = 7x - 3(-2x + 11)$

EUREKA
MATH™

5. $7x - (3x + 5) - 8 = \frac{1}{2}(8x + 20) - 7x + 5$

6. Write at least three equations that have no solution.

Lesson Summary

The distributive property is used to expand expressions. For example, the expression $2(3x - 10)$ is rewritten as $6x - 20$ after the distributive property is applied.

The distributive property is used to simplify expressions. For example, the expression $7x + 11x$ is rewritten as $(7 + 11)x$ and $18x$ after the distributive property is applied.

The distributive property is applied only to terms within a group:

$$4(3x + 5) - 2 = 12x + 20 - 2.$$

Notice that the term -2 is not part of the group and, therefore, not multiplied by 4.

When an equation is transformed into an untrue sentence, such as $5 \neq 11$, we say the equation has *no solution*.

Problem Set

Transform the equation if necessary, and then solve it to find the value of x that makes the equation true.

1. $x - (9x - 10) + 11 = 12x + 3\left(-2x + \dfrac{1}{3}\right)$

2. $7x + 8\left(x + \dfrac{1}{4}\right) = 3(6x - 9) - 8$

3. $-4x - 2(8x + 1) = -(-2x - 10)$

4. $11(x + 10) = 132$

5. $37x + \dfrac{1}{2} - \left(x + \dfrac{1}{4}\right) = 9(4x - 7) + 5$

6. $3(2x - 14) + x = 15 - (-9x - 5)$

7. $8(2x + 9) = 56$

EUREKA
MATH™

Lesson 7: Classification of Solutions

Classwork

Exercises

Solve each of the following equations for x.

1. $7x - 3 = 5x + 5$

2. $7x - 3 = 7x + 5$

3. $7x - 3 = -3 + 7x$

EUREKA
MATH™

Give a brief explanation as to what kind of solution(s) you expect the following linear equations to have. Transform the equations into a simpler form if necessary.

4. $11x - 2x + 15 = 8 + 7 + 9x$

5. $3(x - 14) + 1 = -4x + 5$

6. $-3x + 32 - 7x = -2(5x + 10)$

7. $\frac{1}{2}(8x + 26) = 13 + 4x$

Lesson 7:   Classification of Solutions

EUREKA
MATH

8. Write two equations that have no solutions.

9. Write two equations that have one unique solution each.

10. Write two equations that have infinitely many solutions.

Lesson Summary

There are three classifications of solutions to linear equations: one solution (unique solution), no solution, or infinitely many solutions.

Equations with no solution will, after being simplified, have coefficients of x that are the same on both sides of the equal sign and constants that are different. For example, $x + b = x + c$, where b and c are constants that are not equal. A numeric example is $8x + 5 = 8x - 3$.

Equations with infinitely many solutions will, after being simplified, have coefficients of x and constants that are the same on both sides of the equal sign. For example, $x + a = x + a$, where a is a constant. A numeric example is $6x + 1 = 1 + 6x$.

Problem Set

1. Give a brief explanation as to what kind of solution(s) you expect for the linear equation $18x + \frac{1}{2} = 6(3x + 25)$. Transform the equation into a simpler form if necessary.

2. Give a brief explanation as to what kind of solution(s) you expect for the linear equation $8 - 9x = 15x + 7 + 3x$. Transform the equation into a simpler form if necessary.

3. Give a brief explanation as to what kind of solution(s) you expect for the linear equation $5(x + 9) = 5x + 45$. Transform the equation into a simpler form if necessary.

4. Give three examples of equations where the solution will be unique; that is, only one solution is possible.

5. Solve one of the equations you wrote in Problem 4, and explain why it is the only solution.

6. Give three examples of equations where there will be no solution.

7. Attempt to solve one of the equations you wrote in Problem 6, and explain why it has no solution.

8. Give three examples of equations where there will be infinitely many solutions.

9. Attempt to solve one of the equations you wrote in Problem 8, and explain why it has infinitely many solutions.

Lesson 8: Linear Equations in Disguise

Classwork

Example 3

Can this equation be solved?

$$\frac{6+x}{7x+\frac{2}{3}} = \frac{3}{8}$$

Example 4

Can this equation be solved?

$$\frac{7}{3x+9} = \frac{1}{8}$$

Example 5

In the diagram below, $\triangle ABC \sim \triangle A'B'C'$. Using what we know about similar triangles, we can determine the value of x.

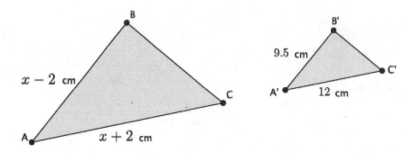

Exercises

Solve the following equations of rational expressions, if possible.

1. $\dfrac{2x+1}{9} = \dfrac{1-x}{6}$

Lesson 8:   Linear Equations in Disguise

EUREKA
MATH

2. $\dfrac{5 + 2x}{3x - 1} = \dfrac{6}{7}$

3. $\dfrac{x + 9}{12} = \dfrac{-2x - \frac{1}{2}}{3}$

4. $\dfrac{8}{3 - 4x} = \dfrac{5}{2x + \frac{1}{4}}$

EUREKA
MATH™

Lesson Summary

Some proportions are linear equations in disguise and are solved the same way we normally solve proportions.

When multiplying a fraction with more than one term in the numerator and/or denominator by a number, put the expressions with more than one term in parentheses so that you remember to use the distributive property when transforming the equation. For example:

$$\frac{x+4}{2x-5} = \frac{3}{5}$$
$$5(x+4) = 3(2x-5).$$

The equation $5(x+4) = 3(2x-5)$ is now clearly a linear equation and can be solved using the properties of equality.

Problem Set

Solve the following equations of rational expressions, if possible. If an equation cannot be solved, explain why.

1. $\dfrac{5}{6x-2} = \dfrac{-1}{x+1}$

2. $\dfrac{4-x}{8} = \dfrac{7x-1}{3}$

3. $\dfrac{3x}{x+2} = \dfrac{5}{9}$

4. $\dfrac{\frac{1}{2}x+6}{3} = \dfrac{x-3}{2}$

5. $\dfrac{7-2x}{6} = \dfrac{x-5}{1}$

6. $\dfrac{2x+5}{2} = \dfrac{3x-2}{6}$

7. $\dfrac{6x+1}{3} = \dfrac{9-x}{7}$

8. $\dfrac{\frac{1}{3}x-8}{12} = \dfrac{-2-x}{15}$

9. $\dfrac{3-x}{1-x} = \dfrac{3}{2}$

10. In the diagram below, $\triangle ABC \sim \triangle A'B'C'$. Determine the lengths of $\overline{AC}$ and $\overline{BC}$.

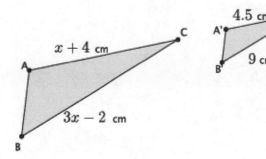

© 2015 Great Minds. eureka-math.org
G8-M3M4M5-SE-B2-1.3.1-01.2016

EUREKA
MATH™

Lesson 9: An Application of Linear Equations

Exercises

1. Write the equation for the 15th step.

2. How many people would see the photo after 15 steps? Use a calculator if needed.

3. Marvin paid an entrance fee of $5 plus an additional $1.25 per game at a local arcade. Altogether, he spent $26.25. Write and solve an equation to determine how many games Marvin played.

4. The sum of four consecutive integers is -26. What are the integers?

5. A book has x pages. How many pages are in the book if Maria read 45 pages of a book on Monday, $\frac{1}{2}$ the book on Tuesday, and the remaining 72 pages on Wednesday?

6. A number increased by 5 and divided by 2 is equal to 75. What is the number?

EUREKA
MATH™

7. The sum of thirteen and twice a number is seven less than six times a number. What is the number?

8. The width of a rectangle is 7 less than twice the length. If the perimeter of the rectangle is 43.6 inches, what is the area of the rectangle?

9. Two hundred fifty tickets for the school dance were sold. On Monday, 35 tickets were sold. An equal number of tickets were sold each day for the next five days. How many tickets were sold on one of those days?

10. Shonna skateboarded for some number of minutes on Monday. On Tuesday, she skateboarded for twice as many minutes as she did on Monday, and on Wednesday, she skateboarded for half the sum of minutes from Monday and Tuesday. Altogether, she skateboarded for a total of three hours. How many minutes did she skateboard each day?

11. In the diagram below, $\triangle ABC \sim \triangle A'B'C'$. Determine the length of $\overline{AC}$ and $\overline{BC}$.

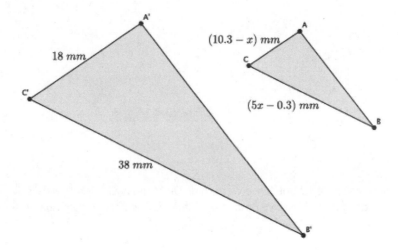

Problem Set

1. You forward an e-card that you found online to three of your friends. They liked it so much that they forwarded it on to four of their friends, who then forwarded it on to four of their friends, and so on. The number of people who saw the e-card is shown below. Let S_1 represent the number of people who saw the e-card after one step, let S_2 represent the number of people who saw the e-card after two steps, and so on.

$$S_1 = 3$$
$$S_2 = 3 + 3 \cdot 4$$
$$S_3 = 3 + 3 \cdot 4 + 3 \cdot 4^2$$
$$S_4 = 3 + 3 \cdot 4 + 3 \cdot 4^2 + 3 \cdot 4^3$$

   a. Find the pattern in the equations.

   b. Assuming the trend continues, how many people will have seen the e-card after 10 steps?

   c. How many people will have seen the e-card after n steps?

For each of the following questions, write an equation, and solve to find each answer.

2. Lisa has a certain amount of money. She spent \$39 and has $\frac{3}{4}$ of the original amount left. How much money did she have originally?

3. The length of a rectangle is 4 more than 3 times the width. If the perimeter of the rectangle is 18.4 cm, what is the area of the rectangle?

4. Eight times the result of subtracting 3 from a number is equal to the number increased by 25. What is the number?

5. Three consecutive odd integers have a sum of 3. What are the numbers?

6. Each month, Liz pays \$35 to her phone company just to use the phone. Each text she sends costs her an additional \$0.05. In March, her phone bill was \$72.60. In April, her phone bill was \$65.85. How many texts did she send each month?

7. Claudia is reading a book that has 360 pages. She read some of the book last week. She plans to read 46 pages today. When she does, she will be $\frac{4}{5}$ of the way through the book. How many pages did she read last week?

8.   In the diagram below, $\triangle ABC \sim \triangle A'B'C'$. Determine the measure of $\angle A$.

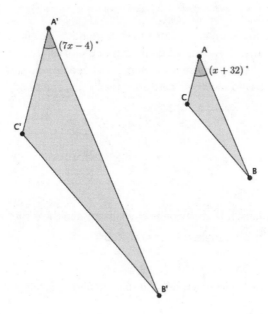

9.   In the diagram below, $\triangle ABC \sim \triangle A'B'C'$. Determine the measure of $\angle A$.

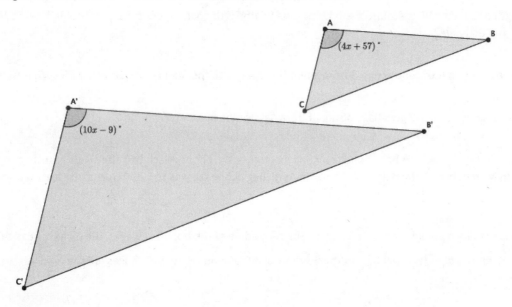

EUREKA
MATH™

Lesson 10: A Critical Look at Proportional Relationships

Classwork

Example 1

Paul walks 2 miles in 25 minutes. How many miles can Paul walk in 137.5 minutes?

Time (in minutes)	Distance (in miles)
25	2

Exercises

1. Wesley walks at a constant speed from his house to school 1.5 miles away. It took him 25 minutes to get to school.

    a.   What fraction represents his constant speed, C?

    b.   You want to know how many miles he has walked after 15 minutes. Let y represent the distance he traveled after 15 minutes of walking at the given constant speed. Write a fraction that represents the constant speed, C, in terms of y.

    c.   Write the fractions from parts (a) and (b) as a proportion, and solve to find how many miles Wesley walked after 15 minutes.

    d.   Let y be the distance in miles that Wesley traveled after x minutes. Write a linear equation in two variables that represents how many miles Wesley walked after x minutes.

2. Stefanie drove at a constant speed from her apartment to her friend's house 20 miles away. It took her 45 minutes to reach her destination.

    a.   What fraction represents her constant speed, C?

EUREKA
MATH

b. What fraction represents constant speed, C, if it takes her x number of minutes to get halfway to her friend's house?

c. Write and solve a proportion using the fractions from parts (a) and (b) to determine how many minutes it takes her to get to the halfway point.

d. Write a two-variable equation to represent how many miles Stefanie can drive over any time interval.

3. The equation that represents how many miles, y, Dave travels after x hours is $y = 50x + 15$. Use the equation to complete the table below.

x (hours)	Linear Equation: $y = 50x + 15$	y (miles)
1	$y = 50(1) + 15$	65
2		
3		
3.5		
4.1		

Lesson 10:   A Critical Look at Proportional Relationships

EUREKA
MATH™

© 2015 Great Minds. eureka-math.org
G8-M3M4M5-SE-B2-1.3.1-01.2016

S.39

Lesson Summary

Average speed is found by taking the total distance traveled in a given time interval, divided by the time interval.

If y is the total distance traveled in a given time interval x, then $\frac{y}{x}$ is the average speed.

If we assume the same average speed over any time interval, then we have constant speed, which can then be used to express a linear equation in two variables relating distance and time.

If $\frac{y}{x} = C$, where C is a constant, then you have constant speed.

Problem Set

1. Eman walks from the store to her friend's house, 2 miles away. It takes her 35 minutes.

   a. What fraction represents her constant speed, C?

   b. Write the fraction that represents her constant speed, C, if she walks y miles in 10 minutes.

   c. Write and solve a proportion using the fractions from parts (a) and (b) to determine how many miles she walks after 10 minutes. Round your answer to the hundredths place.

   d. Write a two-variable equation to represent how many miles Eman can walk over any time interval.

2. Erika drives from school to soccer practice 1.3 miles away. It takes her 7 minutes.

   a. What fraction represents her constant speed, C?

   b. What fraction represents her constant speed, C, if it takes her x minutes to drive exactly 1 mile?

   c. Write and solve a proportion using the fractions from parts (a) and (b) to determine how much time it takes her to drive exactly 1 mile. Round your answer to the tenths place.

   d. Write a two-variable equation to represent how many miles Erika can drive over any time interval.

3. Darla drives at a constant speed of 45 miles per hour.

   a. If she drives for y miles and it takes her x hours, write the two-variable equation to represent the number of miles Darla can drive in x hours.

   b. Darla plans to drive to the market 14 miles from her house, then to the post office 3 miles from the market, and then return home, which is 15 miles from the post office. Assuming she drives at a constant speed the entire time, how much time will she spend driving as she runs her errands? Round your answer to the hundredths place.

4. Aaron walks from his sister's house to his cousin's house, a distance of 4 miles, in 80 minutes. How far does he walk in 30 minutes?

---

5. Carlos walks 4 miles every night for exercise. It takes him exactly 63 minutes to finish his walk.

    a. Assuming he walks at a constant rate, write an equation that represents how many miles, y, Carlos can walk in x minutes.

    b. Use your equation from part (a) to complete the table below. Use a calculator, and round all values to the hundredths place.

x (minutes)	Linear Equation:	y (miles)
15		
30		
40		
60		
75		

This page intentionally left blank

Lesson 11: Constant Rate

Example 1

Pauline mows a lawn at a constant rate. Suppose she mows a 35-square-foot lawn in 2.5 minutes. What area, in square feet, can she mow in 10 minutes? t minutes?

t (time in minutes)	Linear Equation:	y (area in square feet)

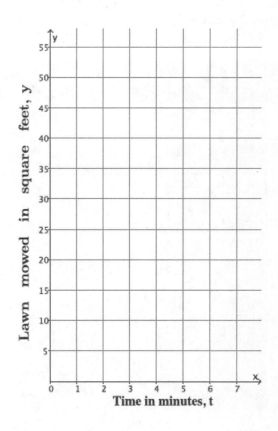

Example 2

Water flows at a constant rate out of a faucet. Suppose the volume of water that comes out in three minutes is 10.5 gallons. How many gallons of water come out of the faucet in t minutes?

t (time in minutes)	Linear Equation:	V (in gallons)
0		
1		
2		
3		
4		

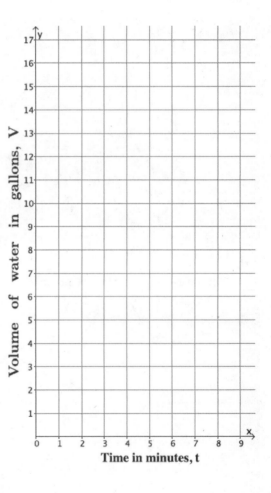

EUREKA MATH™

Exercises

1. Juan types at a constant rate. He can type a full page of text in $3\frac{1}{2}$ minutes. We want to know how many pages, p, Juan can type after t minutes.

    a. Write the linear equation in two variables that represents the number of pages Juan types in any given time interval.

    b. Complete the table below. Use a calculator, and round your answers to the tenths place.

t (time in minutes)	Linear Equation:	p (pages typed)
0		
5		
10		
15		
20		

    c. Graph the data on a coordinate plane.

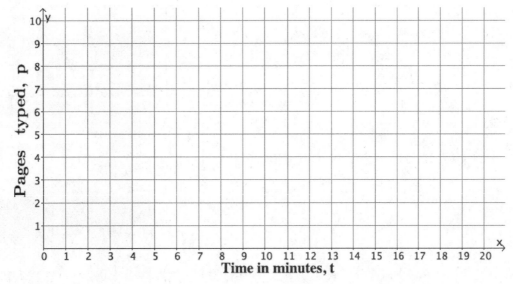

EUREKA
MATH™

    d.   About how long would it take Juan to type a 5-page paper? Explain.

2.   Emily paints at a constant rate. She can paint 32 square feet in 5 minutes. What area, A, in square feet, can she paint in t minutes?

    a.   Write the linear equation in two variables that represents the number of square feet Emily can paint in any given time interval.

    b.   Complete the table below. Use a calculator, and round answers to the tenths place.

t (time in minutes)	Linear Equation:	A (area painted in square feet)
0		
1		
2		
3		
4		

© 2015 Great Minds. eureka-math.org
G8-M3M4M5-SE-B2-1.3.1-01.2016

c.   Graph the data on a coordinate plane.

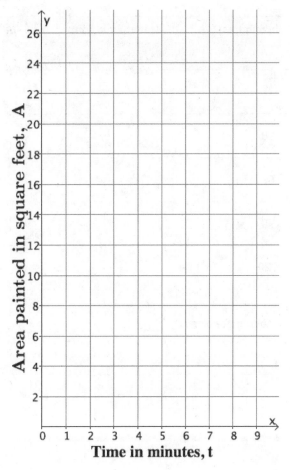

d.   About how many square feet can Emily paint in $2\frac{1}{2}$ minutes? Explain.

3.   Joseph walks at a constant speed. He walked to a store that is one-half mile away in 6 minutes. How many miles, m, can he walk in t minutes?

a.   Write the linear equation in two variables that represents the number of miles Joseph can walk in any given time interval, t.

EUREKA
MATH™

Lesson 11:   Constant Rate

S.47

© 2015 Great Minds. eureka-math.org
G8-M3M4M5-SE-B2-1.3.1-01.2016

b. Complete the table below. Use a calculator, and round answers to the tenths place.

t (time in minutes)	Linear Equation:	m (distance in miles)
0		
30		
60		
90		
120		

c. Graph the data on a coordinate plane.

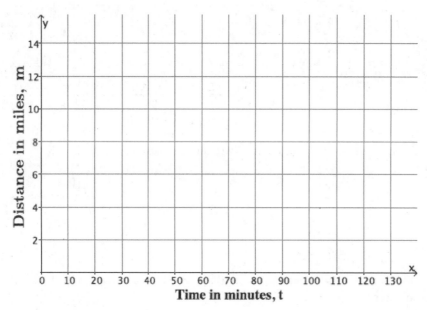

d. Joseph's friend lives 4 miles away from him. About how long would it take Joseph to walk to his friend's house? Explain.

EUREKA
MATH

Lesson Summary

When constant rate is stated for a given problem, then you can express the situation as a two-variable equation. The equation can be used to complete a table of values that can then be graphed on a coordinate plane.

Problem Set

1. A train travels at a constant rate of 45 miles per hour.

    a. What is the distance, d, in miles, that the train travels in t hours?

    b. How many miles will it travel in 2.5 hours?

2. Water is leaking from a faucet at a constant rate of $\frac{1}{3}$ gallon per minute.

    a. What is the amount of water, w, in gallons per minute, that is leaked from the faucet after t minutes?

    b. How much water is leaked after an hour?

3. A car can be assembled on an assembly line in 6 hours. Assume that the cars are assembled at a constant rate.

    a. How many cars, y, can be assembled in t hours?

    b. How many cars can be assembled in a week?

4. A copy machine makes copies at a constant rate. The machine can make 80 copies in $2\frac{1}{2}$ minutes.

    a. Write an equation to represent the number of copies, n, that can be made over any time interval in minutes, t.

    b. Complete the table below.

t (time in minutes)	Linear Equation:	n (number of copies)
0		
0.25		
0.5		
0.75		
1		

---

c.   Graph the data on a coordinate plane.

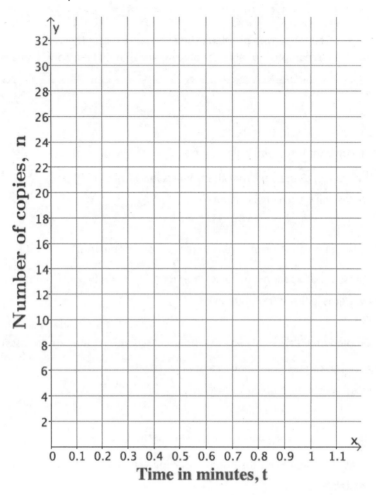

d.   The copy machine runs for 20 seconds and then jams. About how many copies were made before the jam occurred? Explain.

Lesson 11:   Constant Rate

EUREKA
MATH

5. Connor runs at a constant rate. It takes him 34 minutes to run 4 miles.

    a. Write the linear equation in two variables that represents the number of miles Connor can run in any given time interval in minutes, t.

    b. Complete the table below. Use a calculator, and round answers to the tenths place.

t (time in minutes)	Linear Equation:	m (distance in miles)
0		
15		
30		
45		
60		

    c. Graph the data on a coordinate plane.

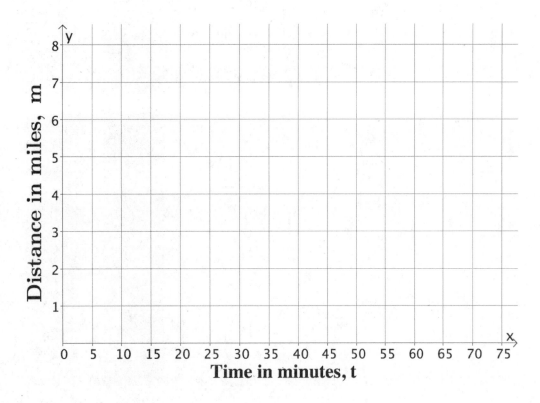

    d. Connor ran for 40 minutes before tripping and spraining his ankle. About how many miles did he run before he had to stop? Explain.

This page intentionally left blank

Lesson 12: Linear Equations in Two Variables

Opening Exercise

Emily tells you that she scored 32 points in a basketball game. Write down all the possible ways she could have scored 32 with only two- and three-point baskets. Use the table below to organize your work.

Number of Two-Pointers	Number of Three-Pointers

Let x be the number of two-pointers and y be the number of three-pointers that Emily scored. Write an equation to represent the situation.

Exploratory Challenge/Exercises

1. Find five solutions for the linear equation $x + y = 3$, and plot the solutions as points on a coordinate plane.

x	Linear Equation: $x + y = 3$	y

2. Find five solutions for the linear equation $2x - y = 10$, and plot the solutions as points on a coordinate plane.

x	Linear Equation: $2x - y = 10$	y

EUREKA
MATH™

3. Find five solutions for the linear equation $x + 5y = 21$, and plot the solutions as points on a coordinate plane.

x	Linear Equation: $x + 5y = 21$	y

4. Consider the linear equation $\frac{2}{5}x + y = 11$.

    a. Will you choose to fix values for x or y? Explain.

    b. Are there specific numbers that would make your computational work easier? Explain.

c.   Find five solutions to the linear equation $\frac{2}{5}x + y = 11$, and plot the solutions as points on a coordinate plane.

x	Linear Equation: $\frac{2}{5}x + y = 11$	y

5.   At the store, you see that you can buy a bag of candy for $2 and a drink for $1. Assume you have a total of $35 to spend. You are feeling generous and want to buy some snacks for you and your friends.

a.   Write an equation in standard form to represent the number of bags of candy, x, and the number of drinks, y, that you can buy with $35.

EUREKA
MATH™

b. Find five solutions to the linear equation from part (a), and plot the solutions as points on a coordinate plane.

x	Linear Equation:	y

Lesson Summary

A linear equation in two-variables x and y is in standard form if it is of the form $ax + by = c$ for numbers a, b, and c, where a and b are both not zero. The numbers a, b, and c are called constants.

A solution to a linear equation in two variables is the ordered pair (x, y) that makes the given equation true. Solutions can be found by fixing a number for x and solving for y or fixing a number for y and solving for x.

Problem Set

1. Consider the linear equation $x - \frac{3}{2}y = -2$.

    a. Will you choose to fix values for x or y? Explain.

    b. Are there specific numbers that would make your computational work easier? Explain.

    c. Find five solutions to the linear equation $x - \frac{3}{2}y = -2$, and plot the solutions as points on a coordinate plane.

x	Linear Equation: $x - \frac{3}{2}y = -2$	y

2. Find five solutions for the linear equation $\frac{1}{3}x + y = 12$, and plot the solutions as points on a coordinate plane.

3. Find five solutions for the linear equation $-x + \frac{3}{4}y = -6$, and plot the solutions as points on a coordinate plane.

4. Find five solutions for the linear equation $2x + y = 5$, and plot the solutions as points on a coordinate plane.

5. Find five solutions for the linear equation $3x - 5y = 15$, and plot the solutions as points on a coordinate plane.

EUREKA MATH™

Lesson 13: The Graph of a Linear Equation in Two Variables

Classwork

Exercises

1. Find at least ten solutions to the linear equation $3x + y = -8$, and plot the points on a coordinate plane.

x	Linear Equation: $3x + y = -8$	y

What shape is the graph of the linear equation taking?

2. Find at least ten solutions to the linear equation $x - 5y = 11$, and plot the points on a coordinate plane.

x	Linear Equation: $x - 5y = 11$	y

What shape is the graph of the linear equation taking?

EUREKA
MATH™

3.   Compare the solutions you found in Exercise 1 with a partner. Add the partner's solutions to your graph.

     Is the prediction you made about the shape of the graph still true? Explain.

4.   Compare the solutions you found in Exercise 2 with a partner. Add the partner's solutions to your graph.

     Is the prediction you made about the shape of the graph still true? Explain.

5.   Joey predicts that the graph of $-x + 2y = 3$ will look like the graph shown below. Do you agree? Explain why or why not.

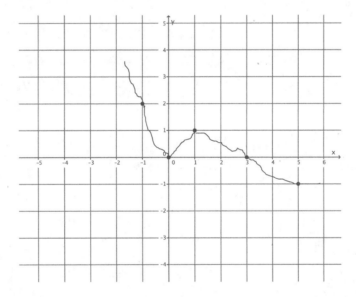

6.   We have looked at some equations that appear to be lines. Can you write an equation that has solutions that do not form a line? Try to come up with one, and prove your assertion on the coordinate plane.

Lesson Summary

One way to determine if a given point is on the graph of a linear equation is by checking to see if it is a solution to the equation. Note that all graphs of linear equations appear to be lines.

Problem Set

1. Find at least ten solutions to the linear equation $\frac{1}{2}x + y = 5$, and plot the points on a coordinate plane.

   What shape is the graph of the linear equation taking?

2. Can the following points be on the graph of the equation $x - y = 0$? Explain.

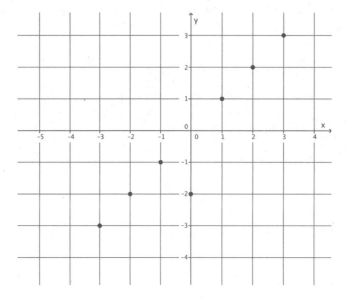

Lesson 13: The Graph of a Linear Equation in Two Variables

EUREKA
MATH™

3.   Can the following points be on the graph of the equation $x + 2y = 2$? Explain.

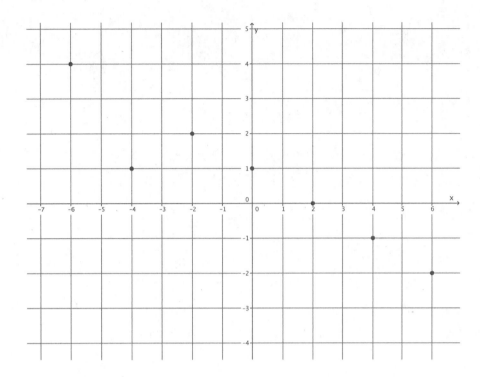

4.   Can the following points be on the graph of the equation $x - y = 7$? Explain.

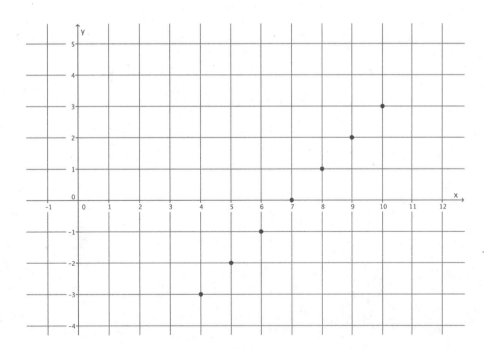

EUREKA
MATH™

5.   Can the following points be on the graph of the equation $x + y = 2$? Explain.

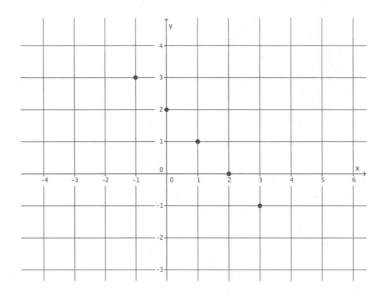

6.   Can the following points be on the graph of the equation $2x - y = 9$? Explain.

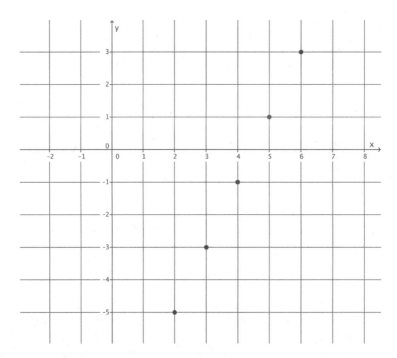

EUREKA
MATH™

7. Can the following points be on the graph of the equation $x - y = 1$? Explain.

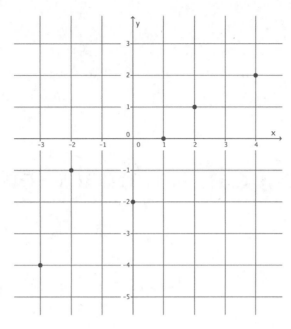

EUREKA
MATH

Lesson 13:   The Graph of a Linear Equation in Two Variables

S.65

© 2015 Great Minds. eureka-math.org
G8-M3M4M5-SE-B2-1.3.1-01.2016

This page intentionally left blank

Lesson 14: The Graph of a Linear Equation—Horizontal and Vertical Lines

Classwork

Exercises

1. Find at least four solutions to graph the linear equation $1x + 2y = 5$.

2. Find at least four solutions to graph the linear equation $1x + 0y = 5$.

3. What was different about the equations in Exercises 1 and 2? What effect did this change have on the graph?

4. Graph the linear equation $x = -2$.

5. Graph the linear equation $x = 3$.

6. What will the graph of $x = 0$ look like?

7. Find at least four solutions to graph the linear equation $2x + 1y = 2$.

8. Find at least four solutions to graph the linear equation $0x + 1y = 2$.

9. What was different about the equations in Exercises 7 and 8? What effect did this change have on the graph?

10. Graph the linear equation $y = -2$.

11. Graph the linear equation $y = 3$.

12. What will the graph of $y = 0$ look like?

Lesson Summary

In a coordinate plane with perpendicular x- and y-axes, a *vertical line* is either the y-axis or any other line parallel to the y-axis. The graph of the linear equation in two variables $ax + by = c$, where $a = 1$ and $b = 0$, is the graph of the equation $x = c$. The graph of $x = c$ is the vertical line that passes through the point $(c, 0)$.

In a coordinate plane with perpendicular x- and y-axes, a *horizontal line* is either the x-axis or any other line parallel to the x-axis. The graph of the linear equation in two variables $ax + by = c$, where $a = 0$ and $b = 1$, is the graph of the equation $y = c$. The graph of $y = c$ is the horizontal line that passes through the point $(0, c)$.

Problem Set

1. Graph the two-variable linear equation $ax + by = c$, where $a = 0$, $b = 1$, and $c = -4$.

2. Graph the two-variable linear equation $ax + by = c$, where $a = 1$, $b = 0$, and $c = 9$.

3. Graph the linear equation $y = 7$.

4. Graph the linear equation $x = 1$.

5. Explain why the graph of a linear equation in the form of $y = c$ is the horizontal line, parallel to the x-axis passing through the point $(0, c)$.

6. Explain why there is only one line with the equation $y = c$ that passes through the point $(0, c)$.

EUREKA
MATH™

Lesson 15: The Slope of a Non-Vertical Line

Opening Exercise

Graph A   **Graph B**

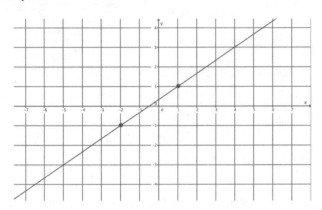   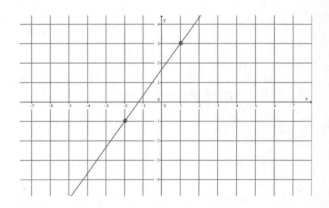

a.   Which graph is steeper?

b.   Write directions that explain how to move from one point on the graph to the other for both Graph A and Graph B.

c.   Write the directions from part (b) as ratios, and then compare the ratios. How does this relate to which graph was steeper in part (a)?

Pair 1:

Graph A   **Graph B**

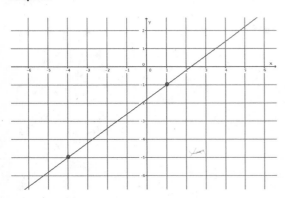   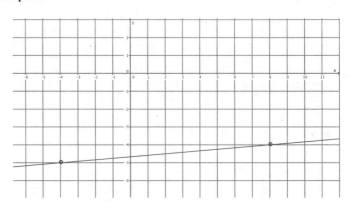

a.   Which graph is steeper?

b.   Write directions that explain how to move from one point on the graph to the other for both Graph A and Graph B.

c.   Write the directions from part (b) as ratios, and then compare the ratios. How does this relate to which graph was steeper in part (a)?

EUREKA MATH

Pair 2:

Graph A

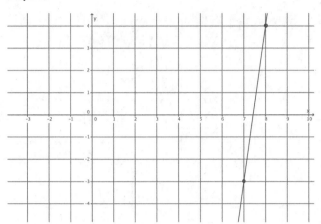

Graph B

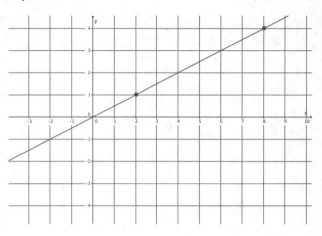

a.   Which graph is steeper?

b.   Write directions that explain how to move from one point on the graph to the other for both Graph A and Graph B.

c.   Write the directions from part (b) as ratios, and then compare the ratios. How does this relate to which graph was steeper in part (a)?

Lesson 15:   The Slope of a Non-Vertical Line

EUREKA
MATH™

© 2015 Great Minds. eureka-math.org
G8-M3M4M5-SE-B2-1.3.1-01.2016

S.71

Pair 3:

Graph A   **Graph B**

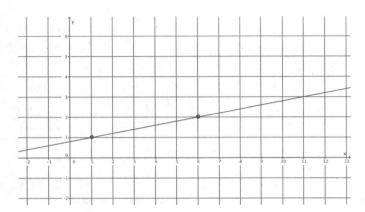   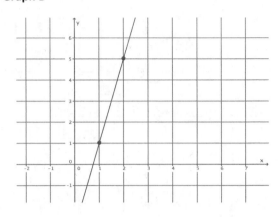

a.   Which graph is steeper?

b.   Write directions that explain how to move from one point on the graph to the other for both Graph A and Graph B.

c.   Write the directions from part (b) as ratios, and then compare the ratios. How does this relate to which graph was steeper in part (a)?

EUREKA
MATH

Pair 4:

Graph A

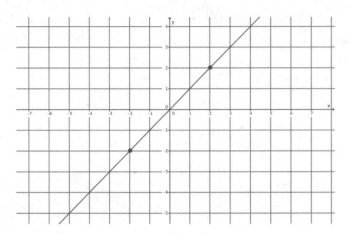

Graph B

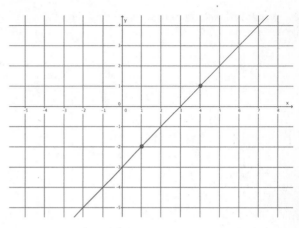

a. Which graph is steeper?

b. Write directions that explain how to move from one point on the graph to the other for both Graph A and Graph B.

c. Write the directions from part (b) as ratios, and then compare the ratios. How does this relate to which graph was steeper in part (a)?

Exercises

Use your transparency to find the slope of each line if needed.

1. What is the slope of this non-vertical line?

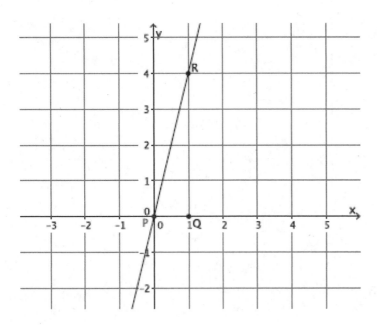

2. What is the slope of this non-vertical line?

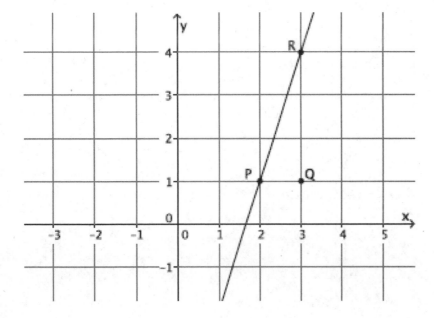

EUREKA
MATH™

3. Which of the lines in Exercises 1 and 2 is steeper? Compare the slopes of each of the lines. Is there a relationship between steepness and slope?

4. What is the slope of this non-vertical line?

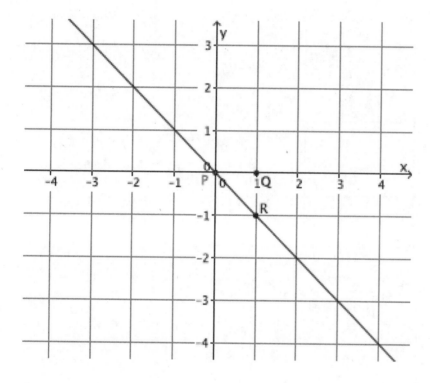

EUREKA
MATH™

Lesson 15:   The Slope of a Non-Vertical Line

© 2015 Great Minds. eureka-math.org
G8-M3M4M5-SE-B2-1.3.1-01.2016

S.75

5.   What is the slope of this non-vertical line?

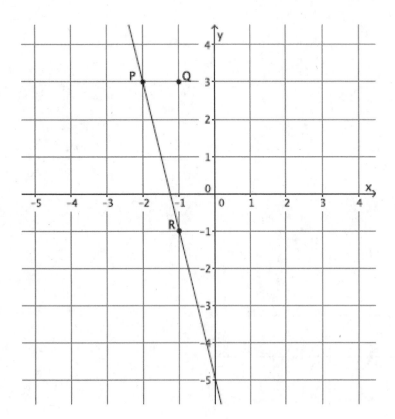

6.   What is the slope of this non-vertical line?

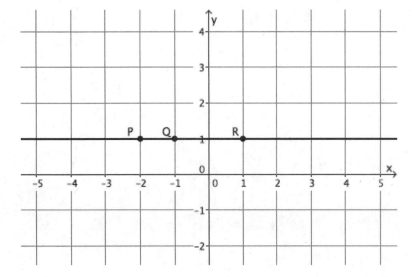

Lesson 15:   The Slope of a Non-Vertical Line

EUREKA
MATH™

Lesson Summary

Slope is a number that can be used to describe the steepness of a line in a coordinate plane. The slope of a line is often represented by the symbol m.

Lines in a coordinate plane that are *left-to-right inclining* have a positive slope, as shown below.

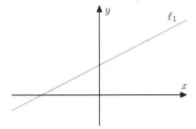

Lines in a coordinate plane that are *left-to-right declining* have a negative slope, as shown below.

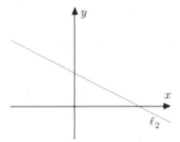

Determine the slope of a line when the horizontal distance between points is fixed at 1 by translating point Q to the origin of the graph and then identifying the y-coordinate of point R; by definition, that number is the slope of the line.

The slope of the line shown below is 2, so $m = 2$, because point R is at 2 on the y-axis.

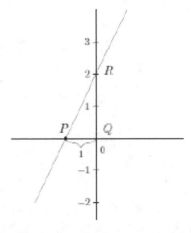

Problem Set

1. Does the graph of the line shown below have a positive or negative slope? Explain.

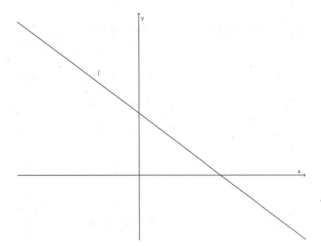

2. Does the graph of the line shown below have a positive or negative slope? Explain.

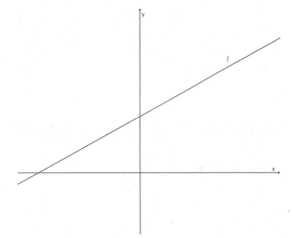

3. What is the slope of this non-vertical line? Use your transparency if needed.

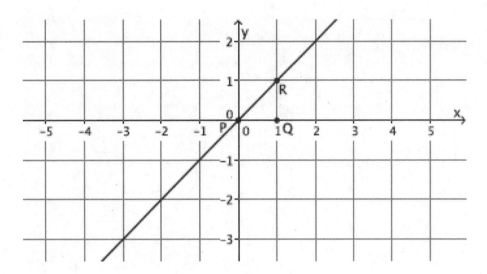

4. What is the slope of this non-vertical line? Use your transparency if needed.

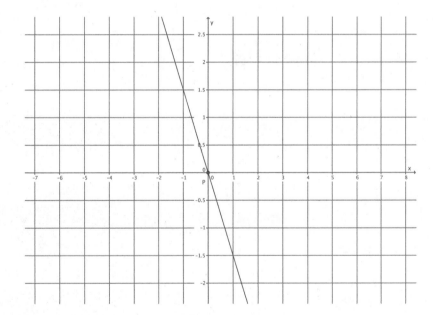

5. What is the slope of this non-vertical line? Use your transparency if needed.

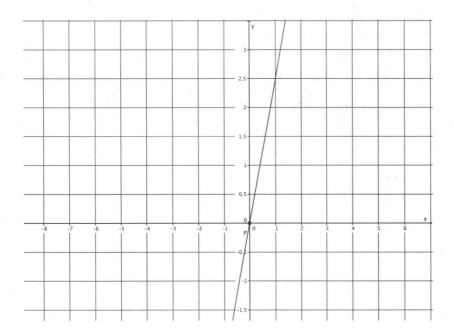

6. What is the slope of this non-vertical line? Use your transparency if needed.

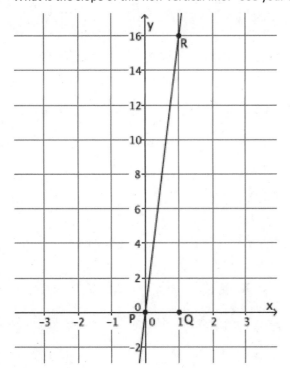

EUREKA
MATH™

7. What is the slope of this non-vertical line? Use your transparency if needed.

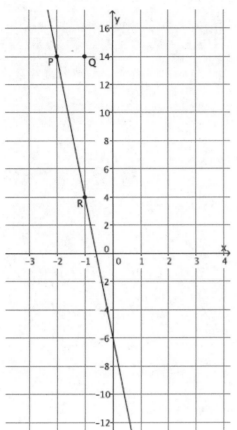

8. What is the slope of this non-vertical line? Use your transparency if needed.

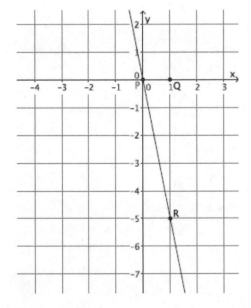

9.   What is the slope of this non-vertical line? Use your transparency if needed.

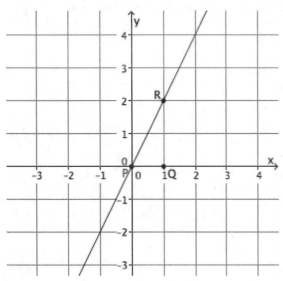

10. What is the slope of this non-vertical line? Use your transparency if needed.

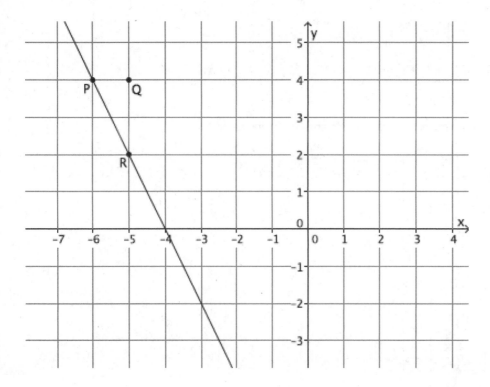

EUREKA
MATH™

11. What is the slope of this non-vertical line? Use your transparency if needed.

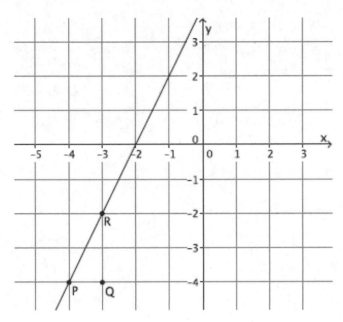

12. What is the slope of this non-vertical line? Use your transparency if needed.

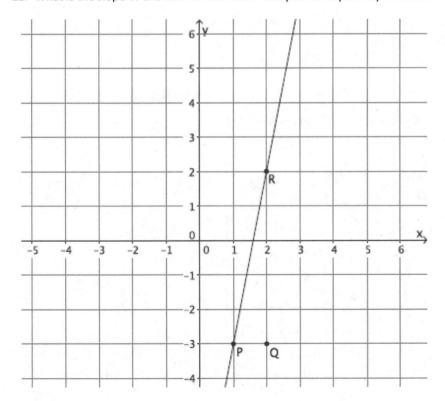

13. What is the slope of this non-vertical line? Use your transparency if needed.

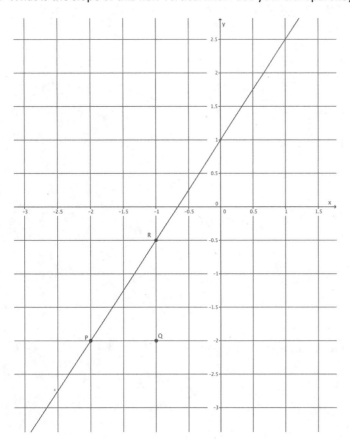

14. What is the slope of this non-vertical line? Use your transparency if needed.

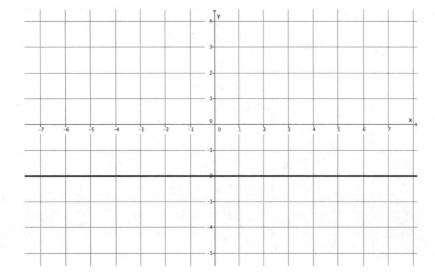

In Lesson 11, you did the work below involving constant rate problems. Use the table and the graphs provided to answer the questions that follow.

15. Suppose the volume of water that comes out in three minutes is 10.5 gallons.

t (time in minutes)	Linear Equation: $V = \dfrac{10.5}{3}t$	V (in gallons)
0	$V = \dfrac{10.5}{3}(0)$	0
1	$V = \dfrac{10.5}{3}(1)$	$\dfrac{10.5}{3} = 3.5$
2	$V = \dfrac{10.5}{3}(2)$	$\dfrac{21}{3} = 7$
3	$V = \dfrac{10.5}{3}(3)$	$\dfrac{31.5}{3} = 10.5$
4	$V = \dfrac{10.5}{3}(4)$	$\dfrac{42}{3} = 14$

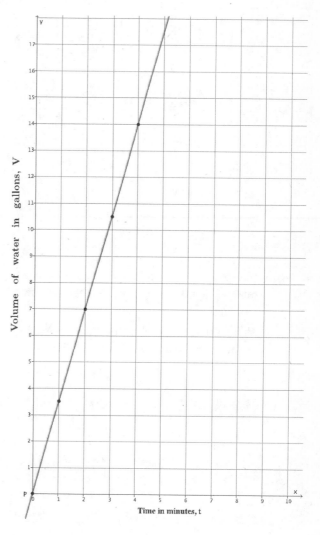

a.   How many gallons of water flow out of the faucet per minute? In other words, what is the unit rate of water flow?

b.   Assume that the graph of the situation is a line, as shown in the graph. What is the slope of the line?

16. Emily paints at a constant rate. She can paint 32 square feet in five minutes.

t (time in minutes)	Linear Equation: $A = \dfrac{32}{5}t$	A (area painted in square feet)
0	$A = \dfrac{32}{5}(0)$	0
1	$A = \dfrac{32}{5}(1)$	$\dfrac{32}{5} = 6.4$
2	$A = \dfrac{32}{5}(2)$	$\dfrac{64}{5} = 12.8$
3	$A = \dfrac{32}{5}(3)$	$\dfrac{96}{5} = 19.2$
4	$A = \dfrac{32}{5}(4)$	$\dfrac{128}{5} = 25.6$

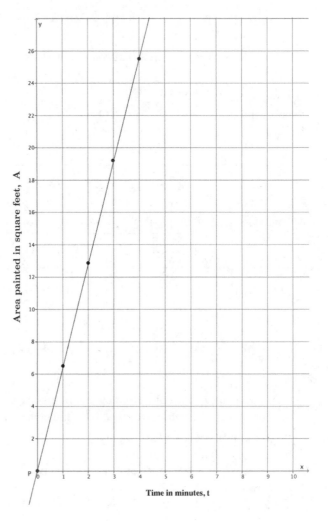

a. How many square feet can Emily paint in one minute? In other words, what is her unit rate of painting?

b. Assume that the graph of the situation is a line, as shown in the graph. What is the slope of the line?

Lesson 15:   The Slope of a Non-Vertical Line

EUREKA MATH

17. A copy machine makes copies at a constant rate. The machine can make 80 copies in $2\frac{1}{2}$ minutes.

t (time in minutes)	Linear Equation: $n = 32t$	n (number of copies)
0	$n = 32(0)$	0
0.25	$n = 32(0.25)$	8
0.5	$n = 32(0.5)$	16
0.75	$n = 32(0.75)$	24
1	$n = 32(1)$	32

a. How many copies can the machine make each minute? In other words, what is the unit rate of the copy machine?

b. Assume that the graph of the situation is a line, as shown in the graph. What is the slope of the line?

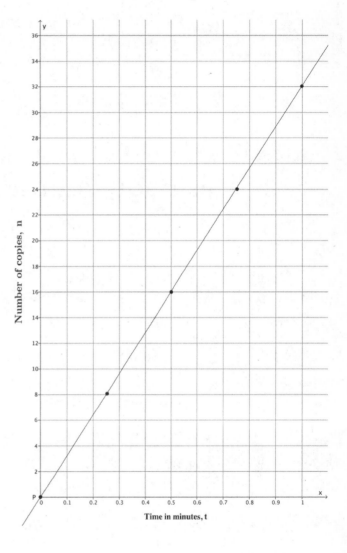

This page intentionally left blank

Lesson 16: The Computation of the Slope of a Non-Vertical Line

Example 1

Using what you learned in the last lesson, determine the slope of the line with the following graph.

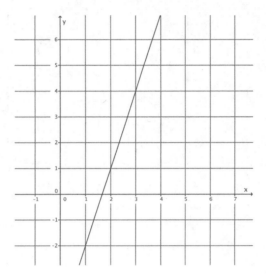

Example 2

Using what you learned in the last lesson, determine the slope of the line with the following graph.

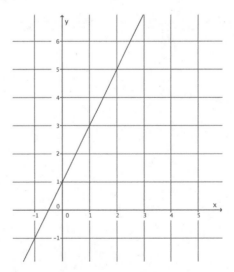

Example 3

What is different about this line compared to the last two examples?

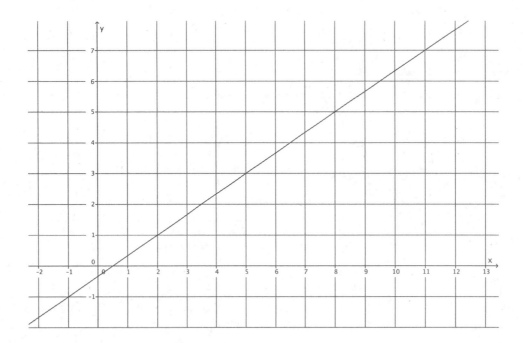

EUREKA
MATH

Exercise

Let's investigate concretely to see if the claim that we can find slope between any two points is true.

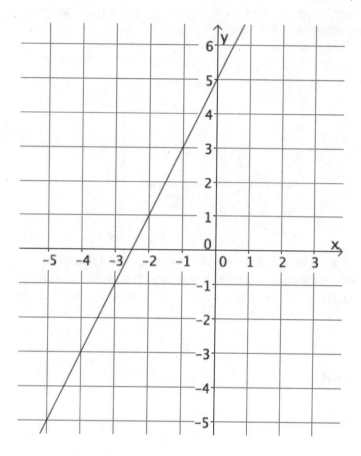

a.   Select any two points on the line to label as P and R.

b.   Identify the coordinates of points P and R.

c.   Find the slope of the line using as many different points as you can. Identify your points, and show your work below.

Lesson Summary

The slope of a line can be calculated using *any* two points on the same line because the slope triangles formed are similar, and corresponding sides will be equal in ratio.

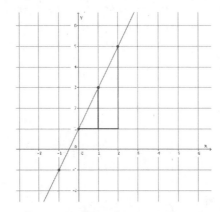

The *slope* of a non-vertical line in a coordinate plane that passes through two different points is the number given by the difference in y-coordinates of those points divided by the difference in the corresponding x-coordinates. For two points $P(p_1, p_2)$ and $R(r_1, r_2)$ on the line where $p_1 \neq r_1$, the slope of the line m can be computed by the formula

$$m = \frac{p_2 - r_2}{p_1 - r_1}.$$

The slope of a vertical line is not defined.

Problem Set

1. Calculate the slope of the line using two different pairs of points.

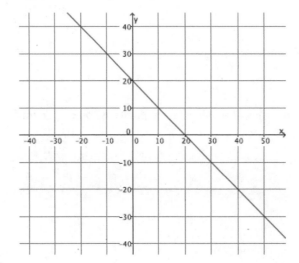

EUREKA
MATH™

2. Calculate the slope of the line using two different pairs of points.

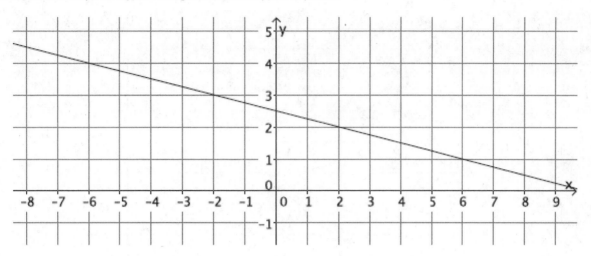

3. Calculate the slope of the line using two different pairs of points.

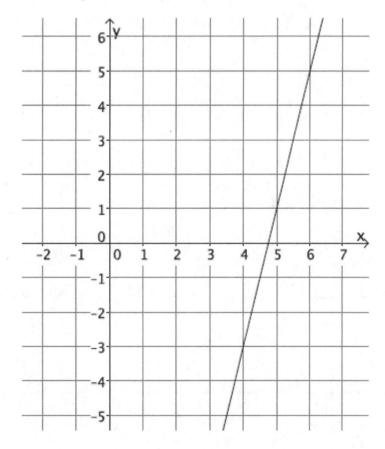

4. Calculate the slope of the line using two different pairs of points.

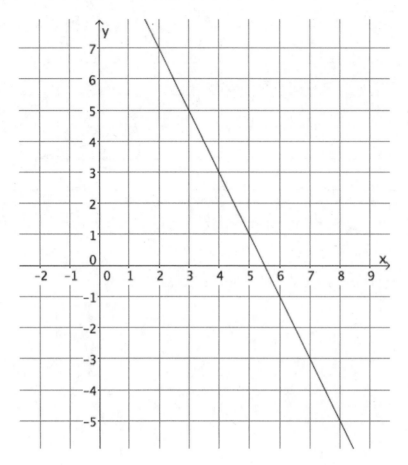

5. Calculate the slope of the line using two different pairs of points.

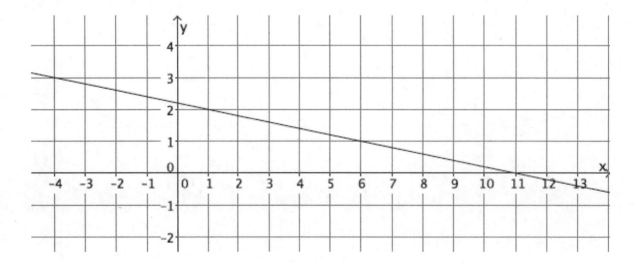

Lesson 16:   The Computation of the Slope of a Non-Vertical Line

EUREKA
MATH™

6. Calculate the slope of the line using two different pairs of points.

   a. Select any two points on the line to compute the slope.

   b. Select two different points on the line to calculate the slope.

   c. What do you notice about your answers in parts (a) and (b)? Explain.

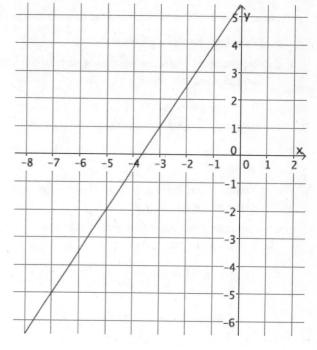

7. Calculate the slope of the line in the graph below.

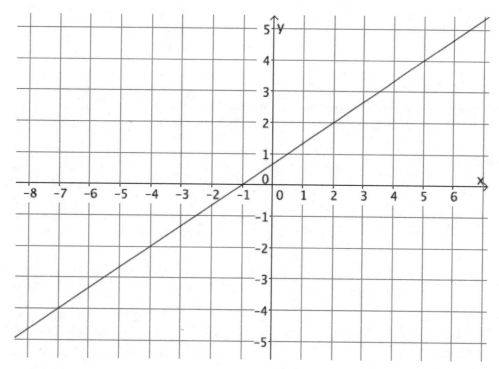

8. Your teacher tells you that a line goes through the points $\left(-6, \frac{1}{2}\right)$ and $(-4,3)$.

    a. Calculate the slope of this line.

    b. Do you think the slope will be the same if the order of the points is reversed? Verify by calculating the slope, and explain your result.

9. Use the graph to complete parts (a)–(c).

    a. Select any two points on the line to calculate the slope.

    b. Compute the slope again, this time reversing the order of the coordinates.

    c. What do you notice about the slopes you computed in parts (a) and (b)?

    d. Why do you think $m = \dfrac{(p_2 - r_2)}{(p_1 - r_1)} = \dfrac{(r_2 - p_2)}{(r_1 - p_1)}$?

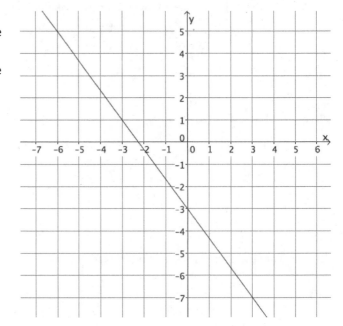

10. Each of the lines in the lesson was non-vertical. Consider the slope of a vertical line, $x = 2$. Select two points on the line to calculate slope. Based on your answer, why do you think the topic of slope focuses only on non-vertical lines?

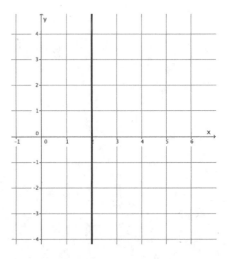

Challenge:

11. A certain line has a slope of $\dfrac{1}{2}$. Name two points that may be on the line.

EUREKA
MATH™

Lesson 17: The Line Joining Two Distinct Points of the Graph $y = mx + b$ Has Slope m

Classwork

Exercises

1. Find at least three solutions to the equation $y = 2x$, and graph the solutions as points on the coordinate plane. Connect the points to make a line. Find the slope of the line.

2. Find at least three solutions to the equation $y = 3x - 1$, and graph the solutions as points on the coordinate plane. Connect the points to make a line. Find the slope of the line.

3. Find at least three solutions to the equation $y = 3x + 1$, and graph the solutions as points on the coordinate plane. Connect the points to make a line. Find the slope of the line.

4. The graph of the equation $y = 7x - 3$ has what slope?

5. The graph of the equation $y = -\frac{3}{4}x - 3$ has what slope?

6. You have $20 in savings at the bank. Each week, you add $2 to your savings. Let y represent the total amount of money you have saved at the end of x weeks. Write an equation to represent this situation, and identify the slope of the equation. What does that number represent?

7. A friend is training for a marathon. She can run 4 miles in 28 minutes. Assume she runs at a constant rate. Write an equation to represent the total distance, y, your friend can run in x minutes. Identify the slope of the equation. What does that number represent?

EUREKA MATH

8. Four boxes of pencils cost \$5. Write an equation that represents the total cost, y, for x boxes of pencils. What is the slope of the equation? What does that number represent?

9. Solve the following equation for y, and then identify the slope of the line: $9x - 3y = 15$.

10. Solve the following equation for y, and then identify the slope of the line: $5x + 9y = 8$.

11. Solve the following equation for y, and then identify the slope of the line: $ax + by = c$.

EUREKA MATH™

Lesson 17:   The Line Joining Two Distinct Points of the Graph
$y = mx + b$ Has Slope m

© 2015 Great Minds. eureka-math.org
G8-M3M4M5-SE-B2-1.3.1-01.2016

S.99

Lesson Summary

The line joining two distinct points of the graph of the linear equation $y = mx + b$ has slope m.

The m of $y = mx + b$ is the number that describes the slope. For example, in the equation $y = -2x + 4$, the slope of the graph of the line is -2.

Problem Set

1. Solve the following equation for y: $-4x + 8y = 24$. Then, answer the questions that follow.

    a. Based on your transformed equation, what is the slope of the linear equation $-4x + 8y = 24$?

    b. Complete the table to find solutions to the linear equation.

x	Transformed Linear Equation:	y

    c. Graph the points on the coordinate plane.

    d. Find the slope between any two points.

    e. The slope you found in part (d) should be equal to the slope you noted in part (a). If so, connect the points to make the line that is the graph of an equation of the form $y = mx + b$ that has slope m.

    f. Note the location (ordered pair) that describes where the line intersects the y-axis.

EUREKA
MATH™

2. Solve the following equation for y: $9x + 3y = 21$. Then, answer the questions that follow.

    a. Based on your transformed equation, what is the slope of the linear equation $9x + 3y = 21$?

    b. Complete the table to find solutions to the linear equation.

x	Transformed Linear Equation:	y

    c. Graph the points on the coordinate plane.

    d. Find the slope between any two points.

    e. The slope you found in part (d) should be equal to the slope you noted in part (a). If so, connect the points to make the line that is the graph of an equation of the form $y = mx + b$ that has slope m.

    f. Note the location (ordered pair) that describes where the line intersects the y-axis.

3. Solve the following equation for y: $2x + 3y = -6$. Then, answer the questions that follow.

    a. Based on your transformed equation, what is the slope of the linear equation $2x + 3y = -6$?

    b. Complete the table to find solutions to the linear equation.

x	Transformed Linear Equation:	y

    c. Graph the points on the coordinate plane.

    d. Find the slope between any two points.

    e. The slope you found in part (d) should be equal to the slope you noted in part (a). If so, connect the points to make the line that is the graph of an equation of the form $y = mx + b$ that has slope m.

    f. Note the location (ordered pair) that describes where the line intersects the y-axis.

4. Solve the following equation for y: $5x - y = 4$. Then, answer the questions that follow.

    a. Based on your transformed equation, what is the slope of the linear equation $5x - y = 4$?

    b. Complete the table to find solutions to the linear equation.

x	Transformed Linear Equation:	y

    c. Graph the points on the coordinate plane.

    d. Find the slope between any two points.

    e. The slope you found in part (d) should be equal to the slope you noted in part (a). If so, connect the points to make the line that is the graph of an equation of the form $y = mx + b$ that has slope m.

    f. Note the location (ordered pair) that describes where the line intersects the y-axis.

Lesson 17:   The Line Joining Two Distinct Points of the Graph $y = mx + b$ Has Slope m

EUREKA
MATH™

Lesson 18: There Is Only One Line Passing Through a Given Point with a Given Slope

Classwork

Opening Exercise

Examine each of the graphs and their equations. Identify the coordinates of the point where the line intersects the y-axis. Describe the relationship between the point and the equation $y = mx + b$.

a. $y = \frac{1}{2}x + 3$

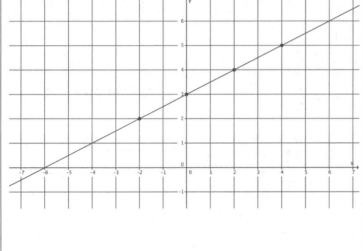

b. $y = -3x + 7$

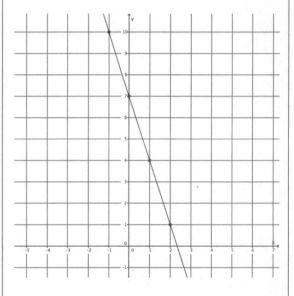

c. $y = -\dfrac{2}{3}x - 2$

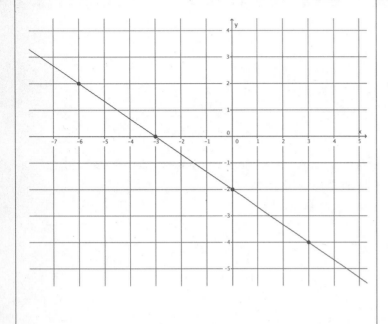

d. $y = 5x - 4$

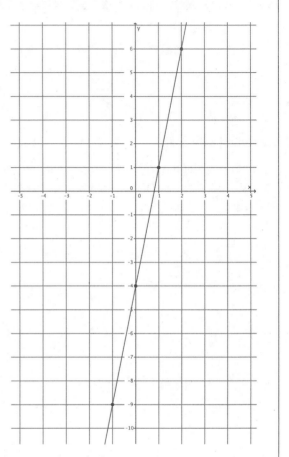

Lesson 18:   There Is Only One Line Passing Through a Given Point with a Given Slope

EUREKA
MATH™

Example 1

Graph the equation $y = \frac{2}{3}x + 1$. Name the slope and y-intercept point.

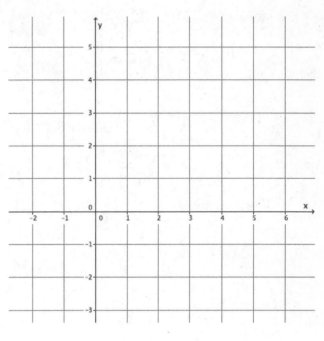

Example 2

Graph the equation $y = -\frac{3}{4}x - 2$. Name the slope and y-intercept point.

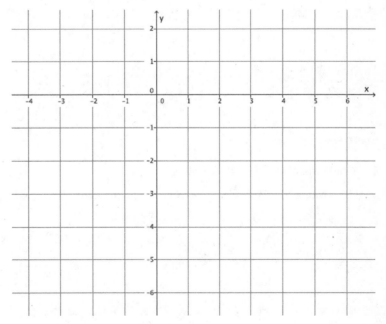

EUREKA
MATH™

Example 3

Graph the equation $y = 4x - 7$. Name the slope and y-intercept point.

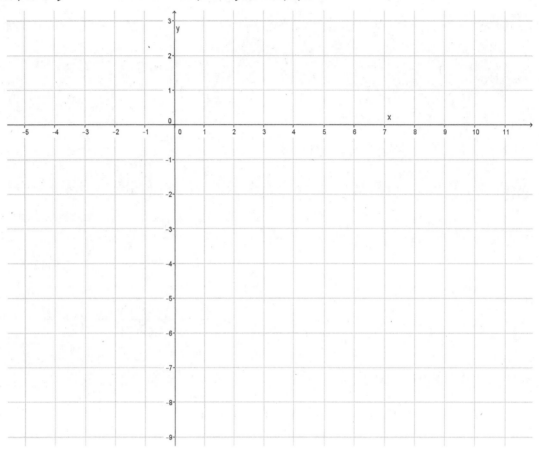

Exercises

1. Graph the equation $y = \frac{5}{2}x - 4$.

    a. Name the slope and the y-intercept point.

Lesson 18:   There Is Only One Line Passing Through a Given Point with a Given Slope

EUREKA
MATH™

b.   Graph the known point, and then use the slope to find a second point before drawing the line.

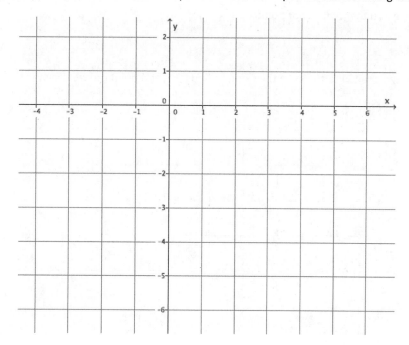

2.   Graph the equation $y = -3x + 6$.

a.   Name the slope and the y-intercept point.

b.   Graph the known point, and then use the slope to find a second point before drawing the line.

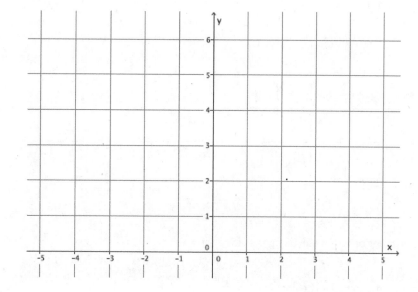

EUREKA
MATH™

Lesson 18:   There Is Only One Line Passing Through a Given Point with a Given Slope

S.107

© 2015 Great Minds. eureka-math.org
G8-M3M4M5-SE-B2-1.3.1-01.2016

3. The equation $y = 1x + 0$ can be simplified to $y = x$. Graph the equation $y = x$.

   a. Name the slope and the y-intercept point.

   b. Graph the known point, and then use the slope to find a second point before drawing the line.

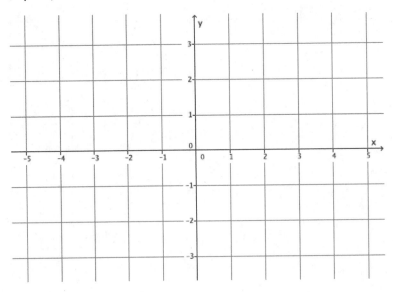

4. Graph the point $(0, 2)$.

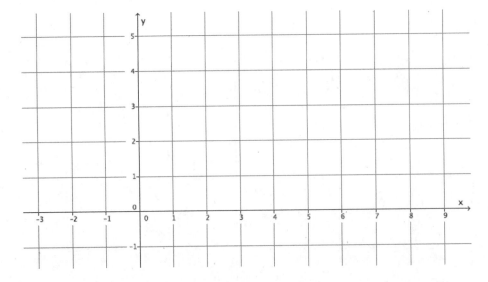

   a. Find another point on the graph using the slope, $m = \frac{2}{7}$.

   b. Connect the points to make the line.

EUREKA
MATH™

c.   Draw a different line that goes through the point $(0, 2)$ with slope $m = \frac{2}{7}$. What do you notice?

5.   A bank put $10 into a savings account when you opened the account. Eight weeks later, you have a total of $24.
     Assume you saved the same amount every week.

   a.   If y is the total amount of money in the savings account and x represents the number of weeks, write an
        equation in the form $y = mx + b$ that describes the situation.

   b.   Identify the slope and the y-intercept point. What do these numbers represent?

   c.   Graph the equation on a coordinate plane.

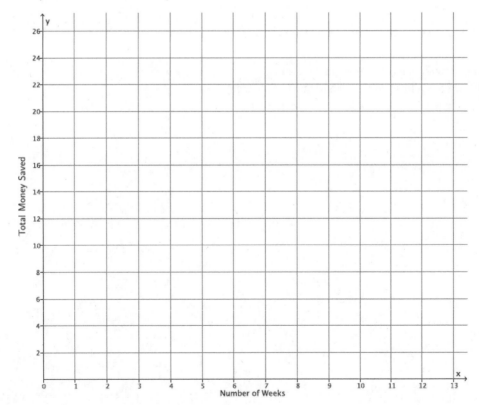

d.   Could any other line represent this situation? For example, could a line through point (0,10) with slope $\frac{7}{5}$ represent the amount of money you save each week? Explain.

6.   A group of friends are on a road trip. After 120 miles, they stop to eat lunch. They continue their trip and drive at a constant rate of 50 miles per hour.

a.   Let y represent the total distance traveled, and let x represent the number of hours driven after lunch. Write an equation to represent the total number of miles driven that day.

b.   Identify the slope and the y-intercept point. What do these numbers represent?

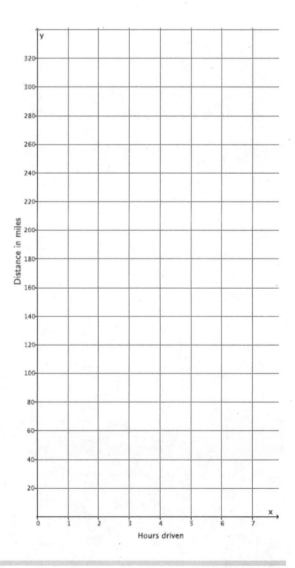

c.   Graph the equation on a coordinate plane.

d.   Could any other line represent this situation? For example, could a line through point $(0, 120)$ with slope 75 represent the total distance the friends drive? Explain.

Lesson 18:   There Is Only One Line Passing Through a Given Point with a Given Slope

EUREKA
MATH

> **Lesson Summary**
>
> The equation $y = mx + b$ is in slope-intercept form. The number m represents the slope of the graph, and the point $(0, b)$ is the location where the graph of the line intersects the y-axis.
>
> To graph a line from the slope-intercept form of a linear equation, begin with the known point, $(0, b)$, and then use the slope to find a second point. Connect the points to graph the equation.
>
> There is only one line passing through a given point with a given slope.

Problem Set

Graph each equation on a separate pair of x- and y-axes.

1. Graph the equation $y = \frac{4}{5}x - 5$.
   a. Name the slope and the y-intercept point.
   b. Graph the known point, and then use the slope to find a second point before drawing the line.

2. Graph the equation $y = x + 3$.
   a. Name the slope and the y-intercept point.
   b. Graph the known point, and then use the slope to find a second point before drawing the line.

3. Graph the equation $y = -\frac{4}{3}x + 4$.
   a. Name the slope and the y-intercept point.
   b. Graph the known point, and then use the slope to find a second point before drawing the line.

4. Graph the equation $y = \frac{5}{2}x$.
   a. Name the slope and the y-intercept point.
   b. Graph the known point, and then use the slope to find a second point before drawing the line.

5. Graph the equation $y = 2x - 6$.
   a. Name the slope and the y-intercept point.
   b. Graph the known point, and then use the slope to find a second point before drawing the line.

6. Graph the equation $y = -5x + 9$.
   a. Name the slope and the y-intercept point.
   b. Graph the known point, and then use the slope to find a second point before drawing the line.

7. Graph the equation $y = \frac{1}{3}x + 1$.

    a. Name the slope and the y-intercept point.

    b. Graph the known point, and then use the slope to find a second point before drawing the line.

8. Graph the equation $5x + 4y = 8$. (Hint: Transform the equation so that it is of the form $y = mx + b$.)

    a. Name the slope and the y-intercept point.

    b. Graph the known point, and then use the slope to find a second point before drawing the line.

9. Graph the equation $-2x + 5y = 30$.

    a. Name the slope and the y-intercept point.

    b. Graph the known point, and then use the slope to find a second point before drawing the line.

10. Let l and l' be two lines with the same slope m passing through the same point P. Show that there is only one line with a slope m, where $m < 0$, passing through the given point P. Draw a diagram if needed.

EUREKA
MATH

Lesson 19: The Graph of a Linear Equation in Two Variables Is a Line

Classwork

Exercises

THEOREM: The graph of a linear equation $y = mx + b$ is a non-vertical line with slope m and passing through $(0, b)$, where b is a constant.

1. Prove the theorem by completing parts (a)–(c). Given two distinct points, P and Q, on the graph of $y = mx + b$, and let l be the line passing through P and Q. You must show the following:

   (1) Any point on the graph of $y = mx + b$ is on line l, and

   (2) Any point on the line l is on the graph of $y = mx + b$.

   a.   Proof of (1): Let R be any point on the graph of $y = mx + b$. Show that R is on l. Begin by assuming it is not. Assume the graph looks like the diagram below where R is on l'.

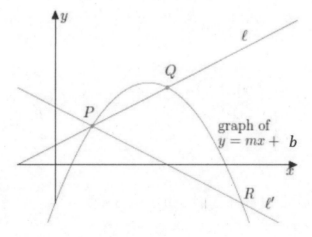

   What is the slope of line l?

What is the slope of line l'?

What can you conclude about lines l and l'? Explain.

b.   Proof of (2): Let S be any point on line l, as shown.

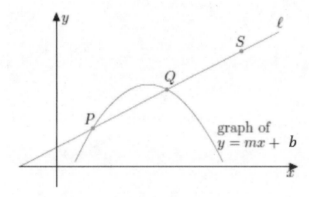

Show that S is a solution to $y = mx + b$. Hint: Use the point $(0, b)$.

EUREKA
MATH

c.   Now that you have shown that any point on the graph of $y = mx + b$ is on line l in part (a), and any point on line l is on the graph of $y = mx + b$ in part (b), what can you conclude about the graphs of linear equations?

2. Use $x = 4$ and $x = -4$ to find two solutions to the equation $x + 2y = 6$. Plot the solutions as points on the coordinate plane, and connect the points to make a line.

a.   Identify two other points on the line with integer coordinates. Verify that they are solutions to the equation $x + 2y = 6$.

b.   When $x = 1$, what is the value of y? Does this solution appear to be a point on the line?

c.   When $x = -3$, what is the value of y? Does this solution appear to be a point on the line?

d.   Is the point $(3, 2)$ on the line?

e.   Is the point $(3, 2)$ a solution to the linear equation $x + 2y = 6$?

3. Use $x = 4$ and $x = 1$ to find two solutions to the equation $3x - y = 9$. Plot the solutions as points on the coordinate plane, and connect the points to make a line.

    a. Identify two other points on the line with integer coordinates. Verify that they are solutions to the equation $3x - y = 9$.

    b. When $x = 4.5$, what is the value of y? Does this solution appear to be a point on the line?

    c. When $x = \frac{1}{2}$, what is the value of y? Does this solution appear to be a point on the line?

    d. Is the point $(2, 4)$ on the line?

    e. Is the point $(2, 4)$ a solution to the linear equation $3x - y = 9$?

4. Use $x = 3$ and $x = -3$ to find two solutions to the equation $2x + 3y = 12$. Plot the solutions as points on the coordinate plane, and connect the points to make a line.

    a. Identify two other points on the line with integer coordinates. Verify that they are solutions to the equation $2x + 3y = 12$.

EUREKA
MATH™

b. When $x = 2$, what is the value of y? Does this solution appear to be a point on the line?

c. When $x = -2$, what is the value of y? Does this solution appear to be a point on the line?

d. Is the point $(8, -3)$ on the line?

e. Is the point $(8, -3)$ a solution to the linear equation $2x + 3y = 12$?

5. Use $x = 4$ and $x = -4$ to find two solutions to the equation $x - 2y = 8$. Plot the solutions as points on the coordinate plane, and connect the points to make a line.

a. Identify two other points on the line with integer coordinates. Verify that they are solutions to the equation $x - 2y = 8$.

b. When $x = 7$, what is the value of y? Does this solution appear to be a point on the line?

EUREKA
MATH™

Lesson 19:   The Graph of a Linear Equation in Two Variables Is a Line

S.117

© 2015 Great Minds. eureka-math.org
G8-M3M4M5-SE-B2-1.3.1-01.2016

   c.   When $x = -3$, what is the value of y? Does this solution appear to be a point on the line?

   d.   Is the point $(-2, -3)$ on the line?

   e.   Is the point $(-2, -3)$ a solution to the linear equation $x - 2y = 8$?

6.   Based on your work in Exercises 2–5, what conclusions can you draw about the points on a line and solutions to a linear equation?

7.   Based on your work in Exercises 2–5, will a point that is not a solution to a linear equation be a point on the graph of a linear equation? Explain.

8.   Based on your work in Exercises 2–5, what conclusions can you draw about the graph of a linear equation?

      Lesson 19:   The Graph of a Linear Equation in Two Variables Is a Line

EUREKA
MATH™

9. Graph the equation $-3x + 8y = 24$ using intercepts.

10. Graph the equation $x - 6y = 15$ using intercepts.

11. Graph the equation $4x + 3y = 21$ using intercepts.

EUREKA
MATH

Lesson 19:   The Graph of a Linear Equation in Two Variables Is a Line

S.119

© 2015 Great Minds. eureka-math.org
G8-M3M4M5-SE-B2-1.3.1-01.2016

Lesson Summary

The graph of a linear equation is a line. A linear equation can be graphed using two-points: the x-intercept point and the y-intercept point.

Example:

Graph the equation: $2x + 3y = 9$.

Replace x with zero, and solve for y to determine the y-intercept point.

$$2(0) + 3y = 9$$
$$3y = 9$$
$$y = 3$$

The y-intercept point is at $(0, 3)$.

Replace y with zero, and solve for x to determine the x-intercept point.

$$2x + 3(0) = 9$$
$$2x = 9$$
$$x = \frac{9}{2}$$

The x-intercept point is at $\left(\frac{9}{2}, 0\right)$.

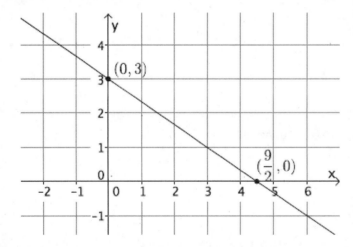

Lesson 19:   The Graph of a Linear Equation in Two Variables Is a Line

EUREKA
MATH™

Problem Set

Graph each of the equations in the Problem Set on a different pair of x- and y-axes.

1. Graph the equation: $y = -6x + 12$.

2. Graph the equation: $9x + 3y = 18$.

3. Graph the equation: $y = 4x + 2$.

4. Graph the equation: $y = -\frac{5}{7}x + 4$.

5. Graph the equation: $\frac{3}{4}x + y = 8$.

6. Graph the equation: $2x - 4y = 12$.

7. Graph the equation: $y = 3$. What is the slope of the graph of this line?

8. Graph the equation: $x = -4$. What is the slope of the graph of this line?

9. Is the graph of $4x + 5y = \frac{3}{7}$ a line? Explain.

10. Is the graph of $6x^2 - 2y = 7$ a line? Explain.

This page intentionally left blank

Lesson 20: Every Line Is a Graph of a Linear Equation

Classwork

Opening Exercise

Figure 1

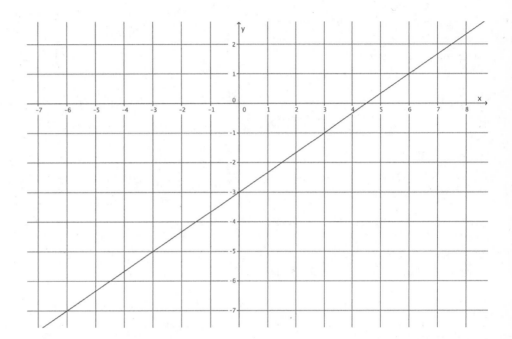

Figure 2

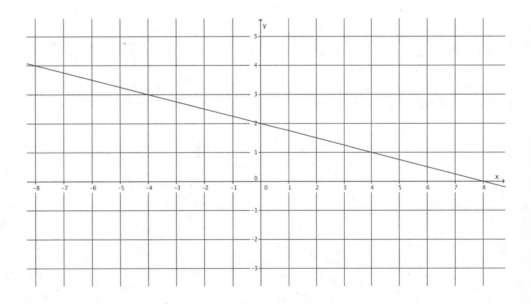

Exercises

1.   Write the equation that represents the
     line shown.

     Use the properties of equality to change
     the equation from slope-intercept form,
     $y = mx + b$, to standard form,
     $ax + by = c$, where a, b, and c are
     integers, and a is not negative.

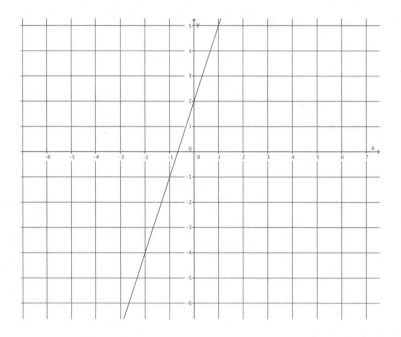

2.   Write the equation that represents the line
     shown.

     Use the properties of equality to change
     the equation from slope-intercept form,
     $y = mx + b$, to standard form,
     $ax + by = c$, where a, b, and c are
     integers, and a is not negative.

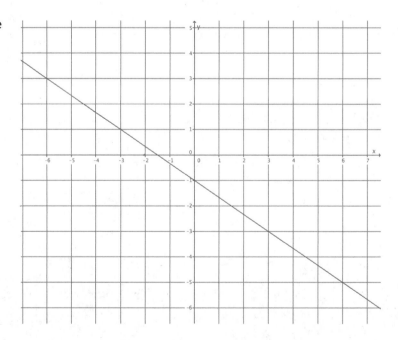

Lesson 20:   Every Line Is a Graph of a Linear Equation

EUREKA
MATH™

3. Write the equation that represents the line shown.

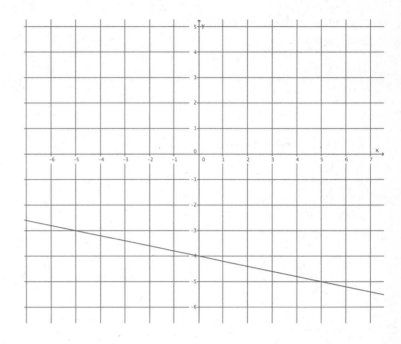

Use the properties of equality to change the equation from slope-intercept form, $y = mx + b$, to standard form, $ax + by = c$, where a, b, and c are integers, and a is not negative.

4. Write the equation that represents the line shown.

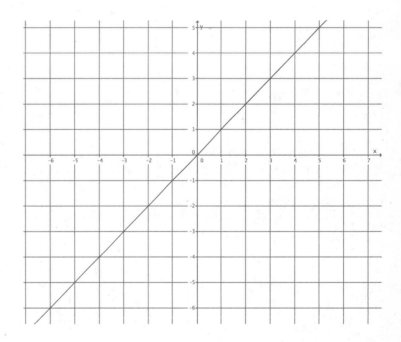

Use the properties of equality to change the equation from slope-intercept form, $y = mx + b$, to standard form, $ax + by = c$, where a, b, and c are integers, and a is not negative.

EUREKA
MATH™

5. Write the equation that represents the
    line shown.

    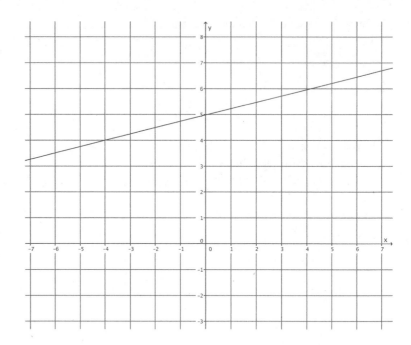

    Use the properties of equality to change
    the equation from slope-intercept form,
    $y = mx + b$, to standard form,
    $ax + by = c$, where a, b, and c are
    integers, and a is not negative.

6. Write the equation that represents the
    line shown.

    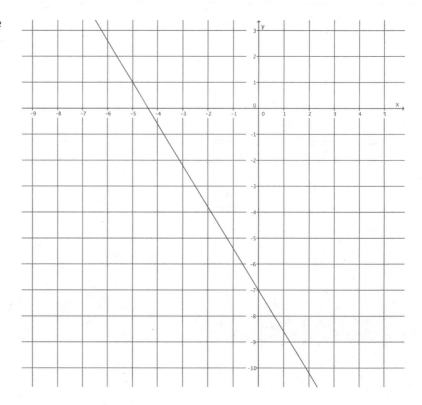

    Use the properties of equality to
    change the equation from slope-
    intercept form, $y = mx + b$, to
    standard form, $ax + by = c$, where
    a, b, and c are integers, and a is not
    negative.

Lesson 20:   Every Line Is a Graph of a Linear Equation

EUREKA
MATH™

Lesson Summary

Write the equation of a line by determining the y-intercept point, $(0, b)$, and the slope, m, and replacing the numbers b and m into the equation $y = mx + b$.

Example:

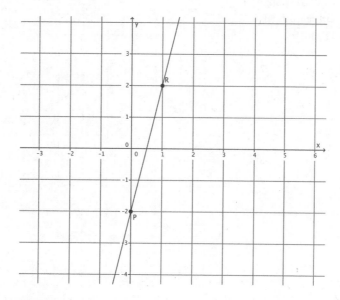

The y-intercept point of this graph is $(0, -2)$.

The slope of this graph is $m = \dfrac{4}{1} = 4$.

The equation that represents the graph of this line is $y = 4x - 2$.

Use the properties of equality to change the equation from slope-intercept form, $y = mx + b$, to standard form, $ax + by = c$, where a, b, and c are integers, and a is not negative.

Problem Set

1. Write the equation that represents the line shown.

    Use the properties of equality to change the equation from slope-intercept form, $y = mx + b$, to standard form, $ax + by = c$, where a, b, and c are integers, and a is not negative.

    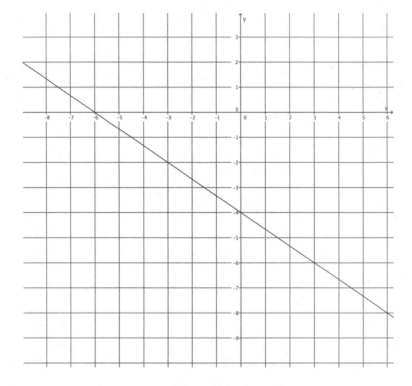

2. Write the equation that represents the line shown.

    Use the properties of equality to change the equation from slope-intercept form, $y = mx + b$, to standard form, $ax + by = c$, where a, b, and c are integers, and a is not negative.

    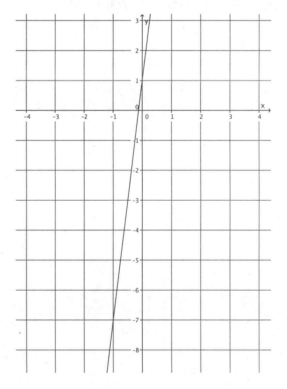

EUREKA
MATH™

3. Write the equation that represents the line shown.

Use the properties of equality to change the equation from slope-intercept form, $y = mx + b$, to standard form, $ax + by = c$, where a, b, and c are integers, and a is not negative.

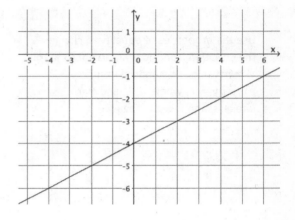

4. Write the equation that represents the line shown.

Use the properties of equality to change the equation from slope-intercept form, $y = mx + b$, to standard form, $ax + by = c$, where a, b, and c are integers, and a is not negative.

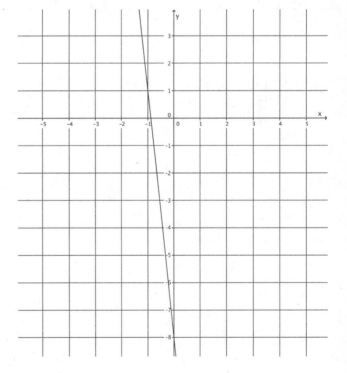

5. Write the equation that represents the line shown.

   Use the properties of equality to change the equation from slope-intercept form, $y = mx + b$, to standard form, $ax + by = c$, where a, b, and c are integers, and a is not negative.

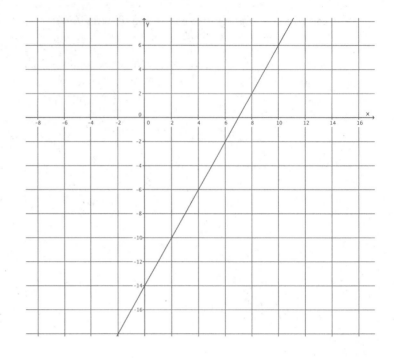

6. Write the equation that represents the line shown.

   Use the properties of equality to change the equation from slope-intercept form, $y = mx + b$, to standard form, $ax + by = c$, where a, b, and c are integers, and a is not negative.

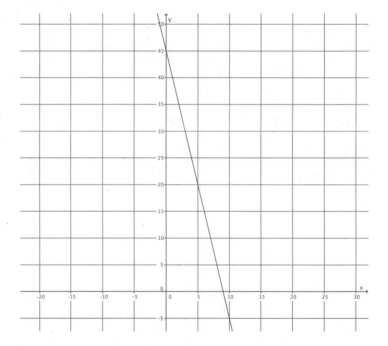

Lesson 21: Some Facts About Graphs of Linear Equations in Two Variables

Example 1

Let a line l be given in the coordinate plane. What linear equation is the graph of line l?

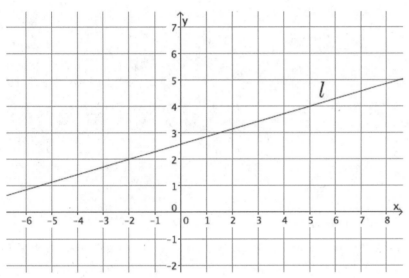

Example 2

Let a line l be given in the coordinate plane. What linear equation is the graph of line l?

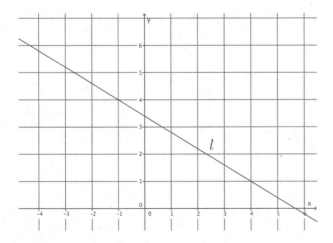

Example 3

Let a line l be given in the coordinate plane. What linear equation is the graph of line l?

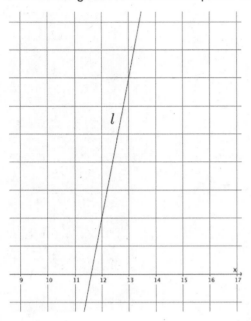

Example 4

Let a line l be given in the coordinate plane. What linear equation is the graph of line l?

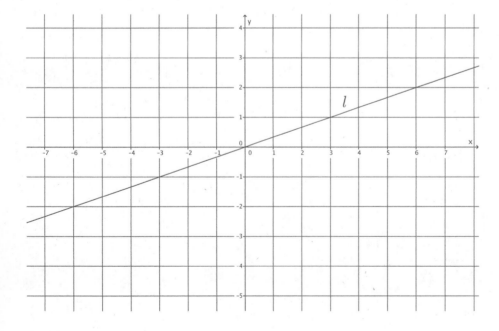

Lesson 21:   Some Facts About Graphs of Linear Equations in Two Variables

EUREKA
MATH™

Exercises

1. Write the equation for the line l shown in the figure.

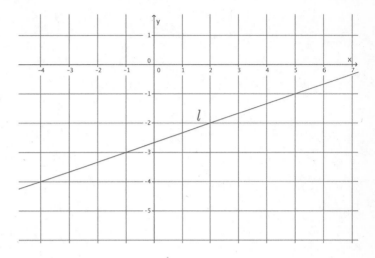

2. Write the equation for the line l shown in the figure.

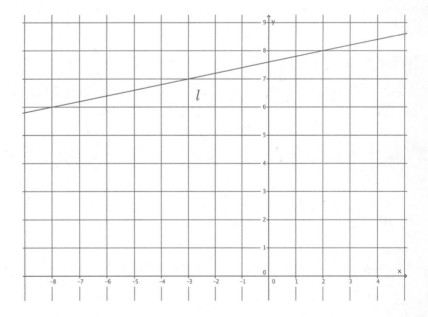

3. Determine the equation of the line that goes through points $(-4, 5)$ and $(2, 3)$.

4. Write the equation for the line l shown in the figure.

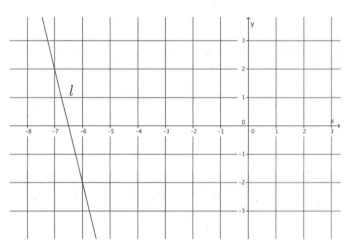

5. A line goes through the point $(8, 3)$ and has slope $m = 4$. Write the equation that represents the line.

Lesson Summary

Let (x_1, y_1) and (x_2, y_2) be the coordinates of two distinct points on a non-vertical line in a coordinate plane. We find the slope of the line by

$$m = \frac{y_2 - y_1}{x_2 - x_1}.$$

This version of the slope formula, using coordinates of x and y instead of p and r, is a commonly accepted version.

As soon as you multiply the slope by the denominator of the fraction above, you get the following equation:

$$m(x_2 - x_1) = y_2 - y_1.$$

This form of an equation is referred to as the *point-slope form* of a linear equation.

Given a known (x, y), then the equation is written as

$$m(x - x_1) = (y - y_1).$$

The following is slope-intercept form of a line:

$$y = mx + b.$$

In this equation, m is slope, and $(0, b)$ is the y-intercept point.

To write the equation of a line, you must have two points, one point and slope, or a graph of the line.

Problem Set

1. Write the equation for the line l shown in the figure.

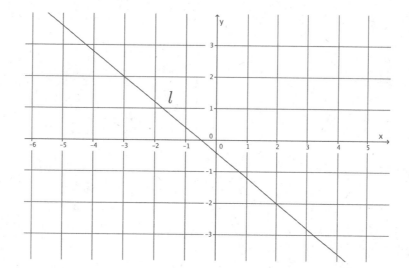

EUREKA
MATH™

2. Write the equation for the line l shown in the figure.

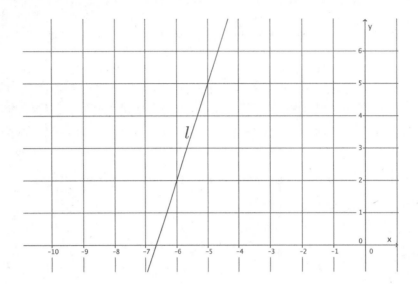

3. Write the equation for the line l shown in the figure.

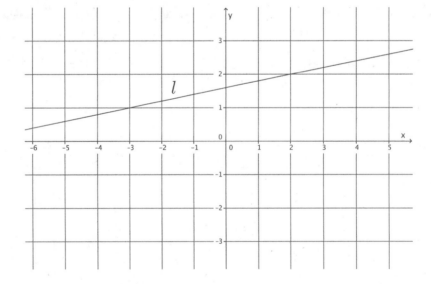

Lesson 21:   Some Facts About Graphs of Linear Equations in Two Variables

EUREKA
MATH™

4. Triangle ABC is made up of line segments formed from the intersection of lines L_{AB}, L_{BC}, and L_{AC}. Write the equations that represent the lines that make up the triangle.

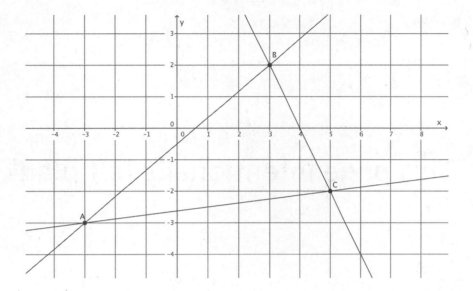

5. Write the equation for the line that goes through point $(-10, 8)$ with slope $m = 6$.

6. Write the equation for the line that goes through point $(12, 15)$ with slope $m = -2$.

7. Write the equation for the line that goes through point $(1, 1)$ with slope $m = -9$.

8. Determine the equation of the line that goes through points $(1, 1)$ and $(3, 7)$.

This page intentionally left blank

Lesson 22: Constant Rates Revisited

Classwork

Exercises

1. Peter paints a wall at a constant rate of 2 square feet per minute. Assume he paints an area y, in square feet, after x minutes.

    a. Express this situation as a linear equation in two variables.

    b. Sketch the graph of the linear equation.

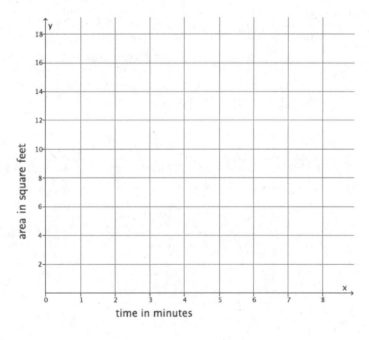

c. Using the graph or the equation, determine the total area he paints after 8 minutes, $1\frac{1}{2}$ hours, and 2 hours. Note that the units are in minutes and hours.

2. The figure below represents Nathan's constant rate of walking.

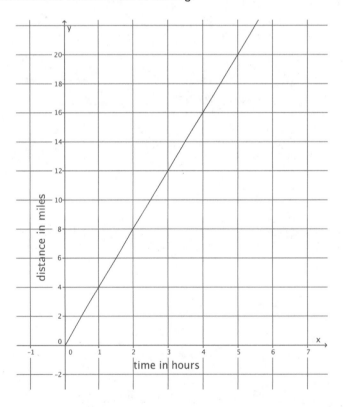

a. Nicole just finished a 5-mile walkathon. It took her 1.4 hours. Assume she walks at a constant rate. Let y represent the distance Nicole walks in x hours. Describe Nicole's walking at a constant rate as a linear equation in two variables.

EUREKA MATH™

b. Who walks at a greater speed? Explain.

3.

a. Susan can type 4 pages of text in 10 minutes. Assuming she types at a constant rate, write the linear equation that represents the situation.

b. The table of values below represents the number of pages that Anne can type, y, in a few selected x minutes. Assume she types at a constant rate.

Minutes (x)	Pages Typed (y)
3	2
5	$\dfrac{10}{3}$
8	$\dfrac{16}{3}$
10	$\dfrac{20}{3}$

Who types faster? Explain.

4.

a. Phil can build 3 birdhouses in 5 days. Assuming he builds birdhouses at a constant rate, write the linear equation that represents the situation.

b. The figure represents Karl's constant rate of building the same kind of birdhouses. Who builds birdhouses faster? Explain.

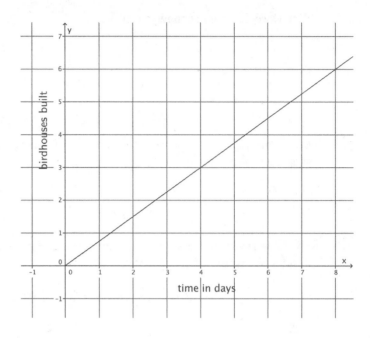

5. Explain your general strategy for comparing proportional relationships.

Lesson 22:   Constant Rates Revisited

EUREKA
MATH™

Lesson Summary

Problems involving constant rate can be expressed as linear equations in two variables.

When given information about two proportional relationships, their rates of change can be compared by comparing the slopes of the graphs of the two proportional relationships.

Problem Set

1.

a. Train A can travel a distance of 500 miles in 8 hours. Assuming the train travels at a constant rate, write the linear equation that represents the situation.

b. The figure represents the constant rate of travel for Train B.

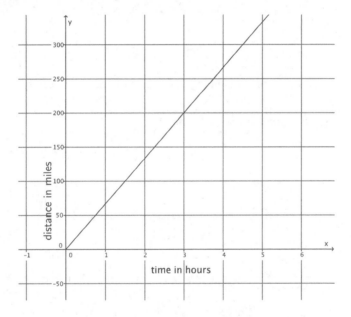

Which train is faster? Explain.

EUREKA
MATH™

Lesson 22:   Constant Rates Revisited

S.143

© 2015 Great Minds. eureka-math.org
G8-M3M4M5-SE-B2-1.3.1-01.2016

2.

   a.   Natalie can paint 40 square feet in 9 minutes. Assuming she paints at a constant rate, write the linear equation that represents the situation.

   b.   The table of values below represents the area painted by Steven for a few selected time intervals. Assume Steven is painting at a constant rate.

Minutes (x)	Area Painted (y)
3	10
5	$\dfrac{50}{3}$
6	20
8	$\dfrac{80}{3}$

     Who paints faster? Explain.

3.

   a.   Bianca can run 5 miles in 41 minutes. Assuming she runs at a constant rate, write the linear equation that represents the situation.

   b.   The figure below represents Cynthia's constant rate of running.

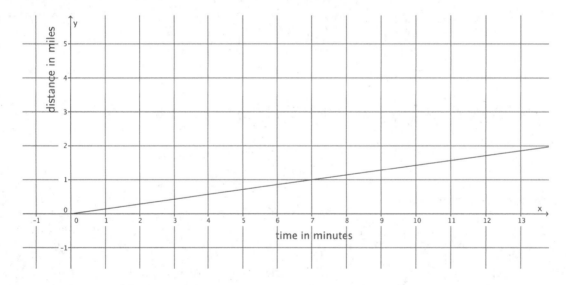

     Who runs faster? Explain.

EUREKA
MATH™

4.

    a.   Geoff can mow an entire lawn of 450 square feet in 30 minutes. Assuming he mows at a constant rate, write the linear equation that represents the situation.

    b.   The figure represents Mark's constant rate of mowing a lawn.

        Who mows faster? Explain.

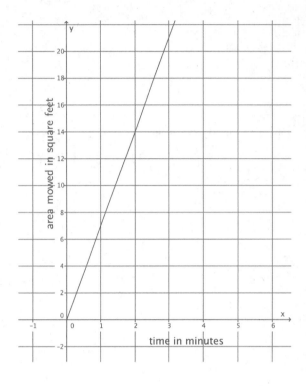

5.

    a.   Juan can walk to school, a distance of 0.75 mile, in 8 minutes. Assuming he walks at a constant rate, write the linear equation that represents the situation.

    b.   The figure below represents Lena's constant rate of walking.

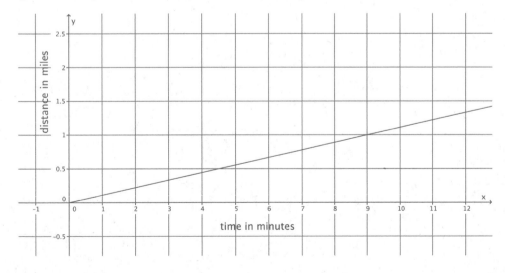

        Who walks faster? Explain.

This page intentionally left blank

Lesson 23: The Defining Equation of a Line

Classwork

Exploratory Challenge/Exercises 1–3

1. Sketch the graph of the equation $9x + 3y = 18$ using intercepts. Then, answer parts (a)–(f) that follow.

    a. Sketch the graph of the equation $y = -3x + 6$ on the same coordinate plane.

    b. What do you notice about the graphs of $9x + 3y = 18$ and $y = -3x + 6$? Why do you think this is so?

    c. Rewrite $y = -3x + 6$ in standard form.

    d. Identify the constants a, b, and c of the equation in standard form from part (c).

e. Identify the constants of the equation $9x + 3y = 18$. Note them as a', b', and c'.

f. What do you notice about $\dfrac{a'}{a}$, $\dfrac{b'}{b}$, and $\dfrac{c'}{c}$?

2. Sketch the graph of the equation $y = \dfrac{1}{2}x + 3$ using the y-intercept point and the slope. Then, answer parts (a)–(f) that follow.

a. Sketch the graph of the equation $4x - 8y = -24$ using intercepts on the same coordinate plane.

b. What do you notice about the graphs of $y = \dfrac{1}{2}x + 3$ and $4x - 8y = -24$? Why do you think this is so?

c. Rewrite $y = \dfrac{1}{2}x + 3$ in standard form.

EUREKA
MATH™

d.   Identify the constants a, b, and c of the equation in standard form from part (c).

e.   Identify the constants of the equation $4x - 8y = -24$. Note them as a', b', and c'.

f.   What do you notice about $\dfrac{a'}{a}$, $\dfrac{b'}{b}$, and $\dfrac{c'}{c}$?

3.   The graphs of the equations $y = \dfrac{2}{3}x - 4$ and $6x - 9y = 36$ are the same line.

a.   Rewrite $y = \dfrac{2}{3}x - 4$ in standard form.

b.   Identify the constants a, b, and c of the equation in standard form from part (a).

c.   Identify the constants of the equation $6x - 9y = 36$. Note them as a', b', and c'.

d.   What do you notice about $\dfrac{a'}{a}$, $\dfrac{b'}{b}$, and $\dfrac{c'}{c}$?

e. You should have noticed that each fraction was equal to the same constant. Multiply that constant by the standard form of the equation from part (a). What do you notice?

Exercises 4–8

4. Write three equations whose graphs are the same line as the equation $3x + 2y = 7$.

5. Write three equations whose graphs are the same line as the equation $x - 9y = \frac{3}{4}$.

EUREKA
MATH™

6. Write three equations whose graphs are the same line as the equation $-9x + 5y = -4$.

7. Write at least two equations in the form $ax + by = c$ whose graphs are the line shown below.

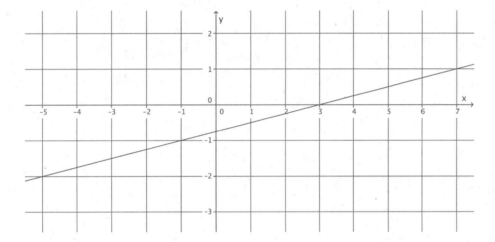

8. Write at least two equations in the form $ax + by = c$ whose graphs are the line shown below.

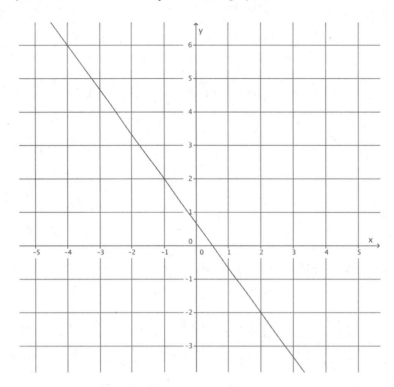

Lesson 23: The Defining Equation of a Line

EUREKA
MATH

Lesson Summary

Two equations define the same line if the graphs of those two equations are the same given line. Two equations that define the same line are the same equation, just in different forms. The equations may look different (different constants, different coefficients, or different forms).

When two equations are written in standard form, $ax + by = c$ and $a'x + b'y = c'$, they define the same line when $\dfrac{a'}{a} = \dfrac{b'}{b} = \dfrac{c'}{c}$ is true.

Problem Set

1. Do the equations $x + y = -2$ and $3x + 3y = -6$ define the same line? Explain.

2. Do the equations $y = -\dfrac{5}{4}x + 2$ and $10x + 8y = 16$ define the same line? Explain.

3. Write an equation that would define the same line as $7x - 2y = 5$.

4. Challenge: Show that if the two lines given by $ax + by = c$ and $a'x + b'y = c'$ are the same when $b = 0$ (vertical lines), then there exists a nonzero number s so that $a' = sa$, $b' = sb$, and $c' = sc$.

5. Challenge: Show that if the two lines given by $ax + by = c$ and $a'x + b'y = c'$ are the same when $a = 0$ (horizontal lines), then there exists a nonzero number s so that $a' = sa$, $b' = sb$, and $c' = sc$.

This page intentionally left blank

Lesson 24: Introduction to Simultaneous Equations

Exercises

1. Derek scored 30 points in the basketball game he played, and not once did he go to the free throw line. That means that Derek scored two-point shots and three-point shots. List as many combinations of two- and three-pointers as you can that would total 30 points.

Number of Two-Pointers	Number of Three-Pointers

Write an equation to describe the data.

2. Derek tells you that the number of two-point shots that he made is five more than the number of three-point shots. How many combinations can you come up with that fit this scenario? (Don't worry about the total number of points.)

Number of Two-Pointers	Number of Three-Pointers

Write an equation to describe the data.

3. Which pair of numbers from your table in Exercise 2 would show Derek's actual score of 30 points?

4. Efrain and Fernie are on a road trip. Each of them drives at a constant speed. Efrain is a safe driver and travels 45 miles per hour for the entire trip. Fernie is not such a safe driver. He drives 70 miles per hour throughout the trip. Fernie and Efrain left from the same location, but Efrain left at 8:00 a.m., and Fernie left at 11:00 a.m. Assuming they take the same route, will Fernie ever catch up to Efrain? If so, approximately when?

   a. Write the linear equation that represents Efrain's constant speed. Make sure to include in your equation the extra time that Efrain was able to travel.

   b. Write the linear equation that represents Fernie's constant speed.

EUREKA
MATH™

c.   Write the system of linear equations that represents this situation.

d.   Sketch the graphs of the two linear equations.

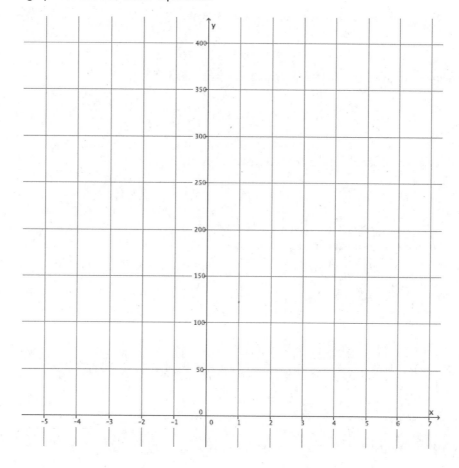

e.   Will Fernie ever catch up to Efrain? If so, approximately when?

f.   At approximately what point do the graphs of the lines intersect?

EUREKA
MATH

Lesson 24:   Introduction to Simultaneous Equations

S.157

© 2015 Great Minds. eureka-math.org
G8-M3M4M5-SE-B2-1.3.1-01.2016

5. Jessica and Karl run at constant speeds. Jessica can run 3 miles in 24 minutes. Karl can run 2 miles in 14 minutes. They decide to race each other. As soon as the race begins, Karl trips and takes 2 minutes to recover.

   a. Write the linear equation that represents Jessica's constant speed. Make sure to include in your equation the extra time that Jessica was able to run.

   b. Write the linear equation that represents Karl's constant speed.

   c. Write the system of linear equations that represents this situation.

   d. Sketch the graphs of the two linear equations.

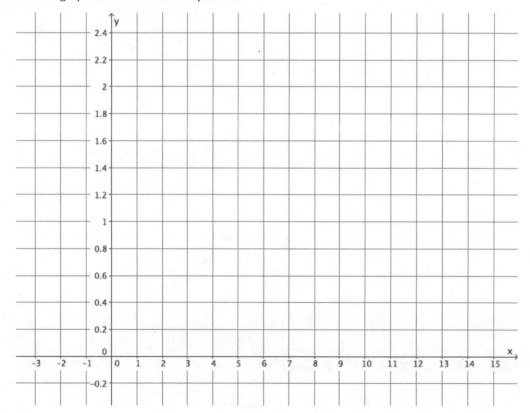

e. Use the graph to answer the questions below.

    i. If Jessica and Karl raced for 3 miles, who would win? Explain.

    ii. At approximately what point would Jessica and Karl be tied? Explain.

Lesson Summary

A *system of linear equations* is a set of two or more linear equations. When graphing a pair of linear equations in two variables, both equations in the system are graphed on the same coordinate plane.

A *solution to a system of two linear equations in two variables* is an ordered pair of numbers that is a solution to both equations. For example, the solution to the system of linear equations $\begin{cases} x + y = 6 \\ x - y = 4 \end{cases}$ is the ordered pair $(5, 1)$ because substituting 5 in for x and 1 in for y results in two true equations: $5 + 1 = 6$ and $5 - 1 = 4$.

Systems of linear equations are notated using brackets to group the equations, for example: $\begin{cases} y = \dfrac{1}{8}x + \dfrac{5}{2} \\ y = \dfrac{4}{25}x \end{cases}$.

Problem Set

1. Jeremy and Gerardo run at constant speeds. Jeremy can run 1 mile in 8 minutes, and Gerardo can run 3 miles in 33 minutes. Jeremy started running 10 minutes after Gerardo. Assuming they run the same path, when will Jeremy catch up to Gerardo?

   a. Write the linear equation that represents Jeremy's constant speed.

   b. Write the linear equation that represents Gerardo's constant speed. Make sure to include in your equation the extra time that Gerardo was able to run.

   c. Write the system of linear equations that represents this situation.

   d. Sketch the graphs of the two equations.

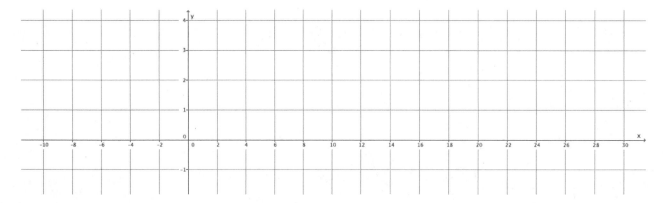

   e. Will Jeremy ever catch up to Gerardo? If so, approximately when?

   f. At approximately what point do the graphs of the lines intersect?

  Introduction to Simultaneous Equations

EUREKA
MATH

2. Two cars drive from town A to town B at constant speeds. The blue car travels 25 miles per hour, and the red car travels 60 miles per hour. The blue car leaves at 9:30 a.m., and the red car leaves at noon. The distance between the two towns is 150 miles.

a. Who will get there first? Write and graph the system of linear equations that represents this situation.

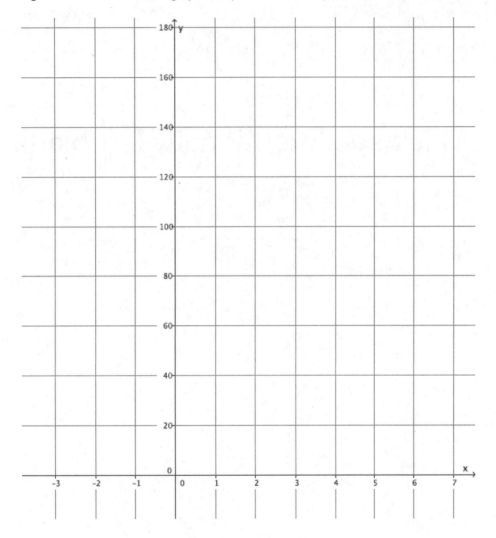

b. At approximately what point do the graphs of the lines intersect?

EUREKA MATH™

This page intentionally left blank

Lesson 25: Geometric Interpretation of the Solutions of a Linear System

Classwork

Exploratory Challenge/Exercises 1–5

1. Sketch the graphs of the linear system on a coordinate plane: $\begin{cases} 2y + x = 12 \\ y = \dfrac{5}{6}x - 2 \end{cases}$.

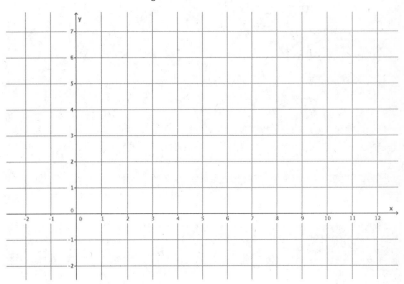

a. Name the ordered pair where the graphs of the two linear equations intersect.

b. Verify that the ordered pair named in part (a) is a solution to $2y + x = 12$.

c. Verify that the ordered pair named in part (a) is a solution to $y = \dfrac{5}{6}x - 2$.

EUREKA
MATH™

d.   Could the point $(4, 4)$ be a solution to the system of linear equations? That is, would $(4, 4)$ make both equations true? Why or why not?

2.   Sketch the graphs of the linear system on a coordinate plane: $\begin{cases} x + y = -2 \\ y = 4x + 3 \end{cases}$.

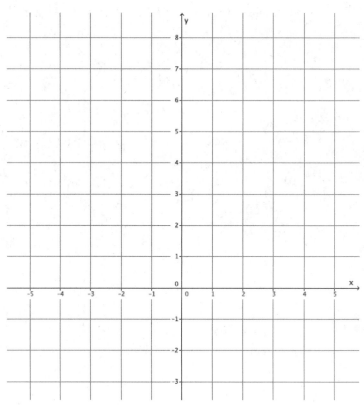

a.   Name the ordered pair where the graphs of the two linear equations intersect.

b.   Verify that the ordered pair named in part (a) is a solution to $x + y = -2$.

EUREKA
MATH™

c. Verify that the ordered pair named in part (a) is a solution to $y = 4x + 3$.

d. Could the point $(-4, 2)$ be a solution to the system of linear equations? That is, would $(-4, 2)$ make both equations true? Why or why not?

3. Sketch the graphs of the linear system on a coordinate plane: $\begin{cases} 3x + y = -3 \\ -2x + y = 2 \end{cases}$.

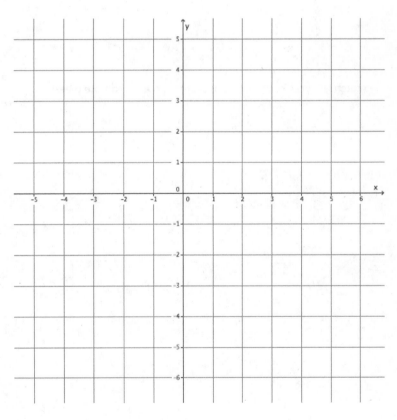

a. Name the ordered pair where the graphs of the two linear equations intersect.

b.   Verify that the ordered pair named in part (a) is a solution to $3x + y = -3$.

c.   Verify that the ordered pair named in part (a) is a solution to $-2x + y = 2$.

d.   Could the point $(1, 4)$ be a solution to the system of linear equations? That is, would $(1, 4)$ make both equations true? Why or why not?

4.   Sketch the graphs of the linear system on a coordinate plane: $\begin{cases} 2x - 3y = 18 \\ 2x + y = 2 \end{cases}$.

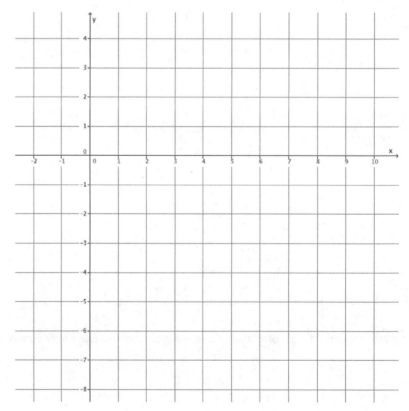

Lesson 25:   Geometric Interpretation of the Solutions of a Linear System

EUREKA
MATH

a.   Name the ordered pair where the graphs of the two linear equations intersect.

b.   Verify that the ordered pair named in part (a) is a solution to $2x - 3y = 18$.

c.   Verify that the ordered pair named in part (a) is a solution to $2x + y = 2$.

d.   Could the point $(3, -1)$ be a solution to the system of linear equations? That is, would $(3, -1)$ make both equations true? Why or why not?

5.   Sketch the graphs of the linear system on a coordinate plane: $\begin{cases} y - x = 3 \\ y = -4x - 2 \end{cases}$.

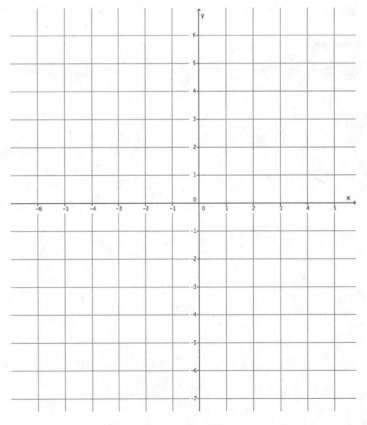

EUREKA
MATH™

a. Name the ordered pair where the graphs of the two linear equations intersect.

b. Verify that the ordered pair named in part (a) is a solution to $y - x = 3$.

c. Verify that the ordered pair named in part (a) is a solution to $y = -4x - 2$.

d. Could the point $(-2, 6)$ be a solution to the system of linear equations? That is, would $(-2, 6)$ make both equations true? Why or why not?

Exercise 6

6. Write two different systems of equations with $(1, -2)$ as the solution.

EUREKA
MATH™

Lesson Summary

When the graphs of a system of linear equations are sketched, and if they are not parallel lines, then the point of intersection of the lines of the graph represents the solution to the system. Two distinct lines intersect at most at one point, if they intersect. The coordinates of that point (x, y) represent values that make both equations of the system true.

Example: The system $\begin{cases} x + y = 3 \\ x - y = 5 \end{cases}$ graphs as shown below.

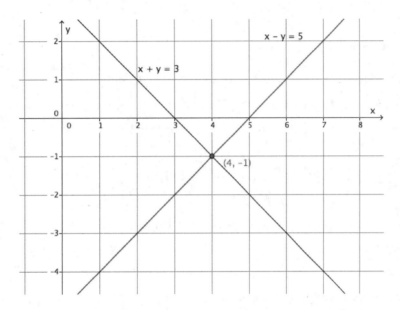

The lines intersect at $(4, -1)$. That means the equations in the system are true when $x = 4$ and $y = -1$.

$$x + y = 3$$
$$4 + (-1) = 3$$
$$3 = 3$$

$$x - y = 5$$
$$4 - (-1) = 5$$
$$5 = 5$$

Problem Set

1. Sketch the graphs of the linear system on a coordinate plane: $\begin{cases} y = \frac{1}{3}x + 1 \\ y = -3x + 11 \end{cases}$.

   a. Name the ordered pair where the graphs of the two linear equations intersect.

   b. Verify that the ordered pair named in part (a) is a solution to $y = \frac{1}{3}x + 1$.

   c. Verify that the ordered pair named in part (a) is a solution to $y = -3x + 11$.

2. Sketch the graphs of the linear system on a coordinate plane: $\begin{cases} y = \frac{1}{2}x + 4 \\ x + 4y = 4 \end{cases}$.

   a. Name the ordered pair where the graphs of the two linear equations intersect.

   b. Verify that the ordered pair named in part (a) is a solution to $y = \frac{1}{2}x + 4$.

   c. Verify that the ordered pair named in part (a) is a solution to $x + 4y = 4$.

3. Sketch the graphs of the linear system on a coordinate plane: $\begin{cases} y = 2 \\ x + 2y = 10 \end{cases}$.

   a. Name the ordered pair where the graphs of the two linear equations intersect.

   b. Verify that the ordered pair named in part (a) is a solution to $y = 2$.

   c. Verify that the ordered pair named in part (a) is a solution to $x + 2y = 10$.

4. Sketch the graphs of the linear system on a coordinate plane: $\begin{cases} -2x + 3y = 18 \\ 2x + 3y = 6 \end{cases}$.

   a. Name the ordered pair where the graphs of the two linear equations intersect.

   b. Verify that the ordered pair named in part (a) is a solution to $-2x + 3y = 18$.

   c. Verify that the ordered pair named in part (a) is a solution to $2x + 3y = 6$.

5. Sketch the graphs of the linear system on a coordinate plane: $\begin{cases} x + 2y = 2 \\ y = \frac{2}{3}x - 6 \end{cases}$.

   a. Name the ordered pair where the graphs of the two linear equations intersect.

   b. Verify that the ordered pair named in part (a) is a solution to $x + 2y = 2$.

   c. Verify that the ordered pair named in part (a) is a solution to $y = \frac{2}{3}x - 6$.

6. Without sketching the graph, name the ordered pair where the graphs of the two linear equations intersect.

$$\begin{cases} x = 2 \\ y = -3 \end{cases}$$

EUREKA
MATH

Lesson 26: Characterization of Parallel Lines

Classwork

Exercises

1. Sketch the graphs of the system. $\begin{cases} y = \dfrac{2}{3}x + 4 \\ y = \dfrac{4}{6}x - 3 \end{cases}$

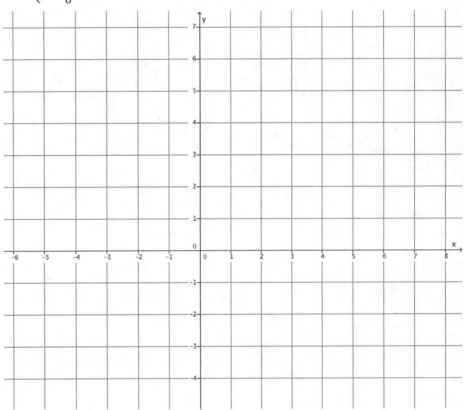

a. Identify the slope of each equation. What do you notice?

b. Identify the y-intercept point of each equation. Are the y-intercept points the same or different?

2. Sketch the graphs of the system. $\begin{cases} y = -\dfrac{5}{4}x + 7 \\ y = -\dfrac{5}{4}x + 2 \end{cases}$

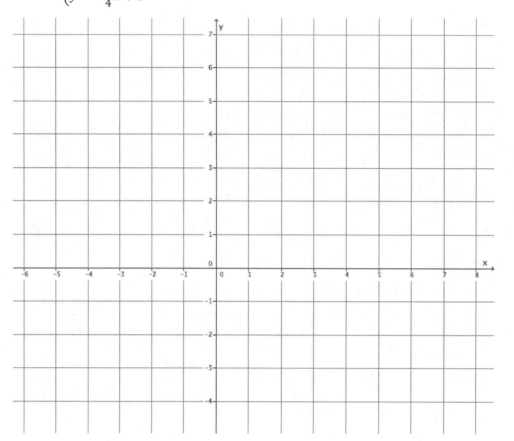

a. Identify the slope of each equation. What do you notice?

b. Identify the y-intercept point of each equation. Are the y-intercept points the same or different?

Lesson 26:   Characterization of Parallel Lines

EUREKA
MATH™

3. Sketch the graphs of the system. $\begin{cases} y = 2x - 5 \\ y = 2x - 1 \end{cases}$

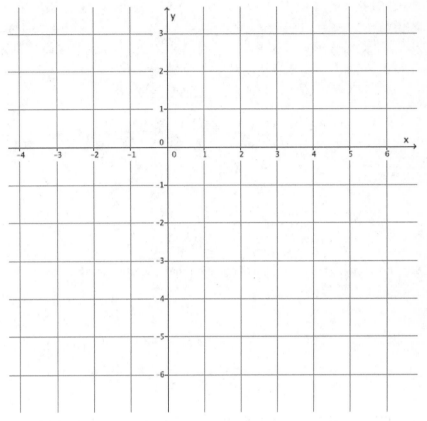

a. Identify the slope of each equation. What do you notice?

b. Identify the y-intercept point of each equation. Are the y-intercept points the same or different?

4. Write a system of equations that has no solution.

5. Write a system of equations that has $(2, 1)$ as a solution.

6. How can you tell if a system of equations has a solution or not?

7. Does the system of linear equations shown below have a solution? Explain.
$$\begin{cases} 6x - 2y = 5 \\ 4x - 3y = 5 \end{cases}$$

8. Does the system of linear equations shown below have a solution? Explain.
$$\begin{cases} -2x + 8y = 14 \\ x = 4y + 1 \end{cases}$$

EUREKA
MATH™

9. Does the system of linear equations shown below have a solution? Explain.

$$\begin{cases} 12x + 3y = -2 \\ 4x + y = 7 \end{cases}$$

10. Genny babysits for two different families. One family pays her $6 each hour and a bonus of $20 at the end of the night. The other family pays her $3 every half hour and a bonus of $25 at the end of the night. Write and solve the system of equations that represents this situation. At what number of hours do the two families pay the same for babysitting services from Genny?

Problem Set

Answer Problems 1–5 without graphing the equations.

1. Does the system of linear equations shown below have a solution? Explain.

$$\begin{cases} 2x + 5y = 9 \\ -4x - 10y = 4 \end{cases}$$

2. Does the system of linear equations shown below have a solution? Explain.

$$\begin{cases} \dfrac{3}{4}x - 3 = y \\ 4x - 3y = 5 \end{cases}$$

3. Does the system of linear equations shown below have a solution? Explain.

$$\begin{cases} x + 7y = 8 \\ 7x - y = -2 \end{cases}$$

4. Does the system of linear equations shown below have a solution? Explain.

$$\begin{cases} y = 5x + 12 \\ 10x - 2y = 1 \end{cases}$$

5. Does the system of linear equations shown below have a solution? Explain.

$$\begin{cases} y = \dfrac{5}{3}x + 15 \\ 5x - 3y = 6 \end{cases}$$

6. Given the graphs of a system of linear equations below, is there a solution to the system that we cannot see on this portion of the coordinate plane? That is, will the lines intersect somewhere on the plane not represented in the picture? Explain.

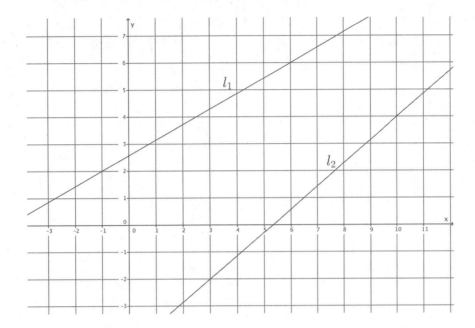

7. Given the graphs of a system of linear equations below, is there a solution to the system that we cannot see on this portion of the coordinate plane? That is, will the lines intersect somewhere on the plane not represented in the picture? Explain.

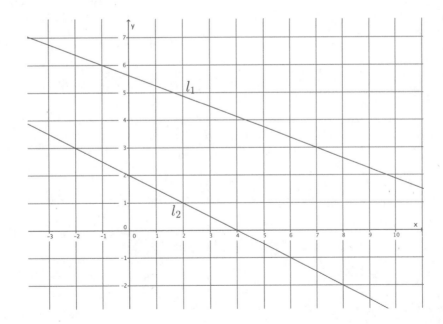

8. Given the graphs of a system of linear equations below, is there a solution to the system that we cannot see on this portion of the coordinate plane? That is, will the lines intersect somewhere on the plane not represented in the picture? Explain.

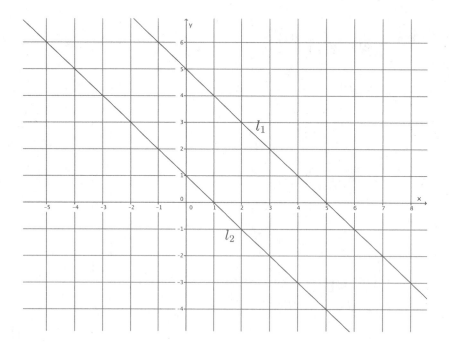

9. Given the graphs of a system of linear equations below, is there a solution to the system that we cannot see on this portion of the coordinate plane? That is, will the lines intersect somewhere on the plane not represented in the picture? Explain.

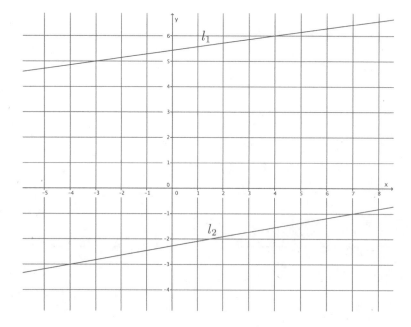

EUREKA
MATH™

10. Given the graphs of a system of linear equations below, is there a solution to the system that we cannot see on this portion of the coordinate plane? That is, will the lines intersect somewhere on the plane not represented in the picture? Explain.

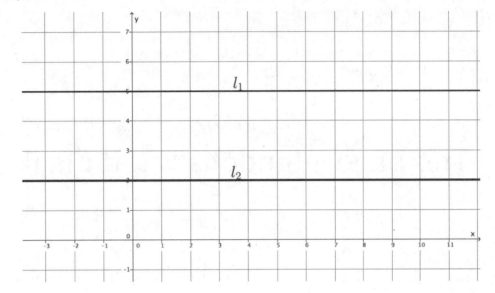

EUREKA
MATH™

This page intentionally left blank

Lesson 27: Nature of Solutions of a System of Linear Equations

Classwork

Exercises

Determine the nature of the solution to each system of linear equations.

1. $\begin{cases} 3x + 4y = 5 \\ y = -\dfrac{3}{4}x + 1 \end{cases}$

2. $\begin{cases} 7x + 2y = -4 \\ x - y = 5 \end{cases}$

3. $\begin{cases} 9x + 6y = 3 \\ 3x + 2y = 1 \end{cases}$

EUREKA
MATH™

Determine the nature of the solution to each system of linear equations. If the system has a solution, find it algebraically, and then verify that your solution is correct by graphing.

4. $\begin{cases} 3x + 3y = -21 \\ x + y = -7 \end{cases}$

5. $\begin{cases} y = \dfrac{3}{2}x - 1 \\ 3y = x + 2 \end{cases}$

EUREKA
MATH™

6. $\begin{cases} x = 12y - 4 \\ x = 9y + 7 \end{cases}$

7. Write a system of equations with $(4, -5)$ as its solution.

Lesson Summary

A system of linear equations can have a unique solution, no solution, or infinitely many solutions.

Systems with a unique solution are comprised of two linear equations whose graphs have different slopes; that is, their graphs in a coordinate plane will be two distinct lines that intersect at only one point.

Systems with no solutions are comprised of two linear equations whose graphs have the same slope but different y-intercept points; that is, their graphs in a coordinate plane will be two parallel lines (with no intersection).

Systems with infinitely many solutions are comprised of two linear equations whose graphs have the same slope and the same y-intercept point; that is, their graphs in a coordinate plane are the same line (i.e., every solution to one equation will be a solution to the other equation).

A system of linear equations can be solved using a substitution method. That is, if two expressions are equal to the same value, then they can be written equal to one another.

Example:

$$\begin{cases} y = 5x - 8 \\ y = 6x + 3 \end{cases}$$

Since both equations in the system are equal to y, we can write the equation $5x - 8 = 6x + 3$ and use it to solve for x and then the system.

Example:

$$\begin{cases} 3x = 4y + 2 \\ x = y + 5 \end{cases}$$

Multiply each term of the equation $x = y + 5$ by 3 to produce the equivalent equation $3x = 3y + 15$. As in the previous example, since both equations equal $3x$, we can write $4y + 2 = 3y + 15$. This equation can be used to solve for y and then the system.

Problem Set

Determine the nature of the solution to each system of linear equations. If the system has a solution, find it algebraically, and then verify that your solution is correct by graphing.

1. $\begin{cases} y = \dfrac{3}{7}x - 8 \\ 3x - 7y = 1 \end{cases}$

2. $\begin{cases} 2x - 5 = y \\ -3x - 1 = 2y \end{cases}$

3. $\begin{cases} x = 6y + 7 \\ x = 10y + 2 \end{cases}$

EUREKA
MATH™

4. $\begin{cases} 5y = \dfrac{15}{4}x + 25 \\ y = \dfrac{3}{4}x + 5 \end{cases}$

5. $\begin{cases} x + 9 = y \\ x = 4y - 6 \end{cases}$

6. $\begin{cases} 3y = 5x - 15 \\ 3y = 13x - 2 \end{cases}$

7. $\begin{cases} 6x - 7y = \dfrac{1}{2} \\ 12x - 14y = 1 \end{cases}$

8. $\begin{cases} 5x - 2y = 6 \\ -10x + 4y = -14 \end{cases}$

9. $\begin{cases} y = \dfrac{3}{2}x - 6 \\ 2y = 7 - 4x \end{cases}$

10. $\begin{cases} 7x - 10 = y \\ y = 5x + 12 \end{cases}$

11. Write a system of linear equations with $(-3, 9)$ as its solution.

This page intentionally left blank

Lesson 28: Another Computational Method of Solving a Linear System

Classwork

Example 1

Use what you noticed about adding equivalent expressions to solve the following system by elimination:

$$\begin{cases} 6x - 5y = 21 \\ 2x + 5y = -5 \end{cases}$$

Example 2

Solve the following system by elimination:

$$\begin{cases} -2x + 7y = 5 \\ 4x - 2y = 14 \end{cases}$$

Example 3

Solve the following system by elimination:

$$\begin{cases} 7x - 5y = -2 \\ 3x - 3y = 7 \end{cases}$$

Exercises

Each of the following systems has a solution. Determine the solution to the system by eliminating one of the variables. Verify the solution using the graph of the system.

1. $\begin{cases} 6x - 7y = -10 \\ 3x + 7y = -8 \end{cases}$

EUREKA
MATH™

2. $\begin{cases} x - 4y = 7 \\ 5x + 9y = 6 \end{cases}$

3. $\begin{cases} 2x - 3y = -5 \\ 3x + 5y = 1 \end{cases}$

Lesson Summary

Systems of linear equations can be solved by eliminating one of the variables from the system. One way to eliminate a variable is by setting both equations equal to the same variable and then writing the expressions equal to one another.

Example: Solve the system $\begin{cases} y = 3x - 4 \\ y = 2x + 1 \end{cases}$.

Since the expressions $3x - 4$ and $2x + 1$ are both equal to y, they can be set equal to each other and the new equation can be solved for x:

$$3x - 4 = 2x + 1$$

Another way to eliminate a variable is by multiplying each term of an equation by the same constant to make an equivalent equation. Then, use the equivalent equation to eliminate one of the variables and solve the system.

Example: Solve the system $\begin{cases} 2x + y = 8 \\ x + y = 10 \end{cases}$.

Multiply the second equation by -2 to eliminate the x.

$$-2(x + y = 10)$$
$$-2x - 2y = -20$$

Now we have the system $\begin{cases} 2x + y = 8 \\ -2x - 2y = -20 \end{cases}$.

When the equations are added together, the x is eliminated.

$$2x + y - 2x - 2y = 8 + (-20)$$
$$y - 2y = 8 + (-20)$$

Once a solution has been found, verify the solution graphically or by substitution.

Problem Set

Determine the solution, if it exists, for each system of linear equations. Verify your solution on the coordinate plane.

1. $\begin{cases} \frac{1}{2}x + 5 = y \\ 2x + y = 1 \end{cases}$

2. $\begin{cases} 9x + 2y = 9 \\ -3x + y = 2 \end{cases}$

3. $\begin{cases} y = 2x - 2 \\ 2y = 4x - 4 \end{cases}$

Lesson 28: Another Computational Method of Solving a Linear System

EUREKA MATH

4. $\begin{cases} 8x + 5y = 19 \\ -8x + y = -1 \end{cases}$

5. $\begin{cases} x + 3 = y \\ 3x + 4y = 7 \end{cases}$

6. $\begin{cases} y = 3x + 2 \\ 4y = 12 + 12x \end{cases}$

7. $\begin{cases} 4x - 3y = 16 \\ -2x + 4y = -2 \end{cases}$

8. $\begin{cases} 2x + 2y = 4 \\ 12 - 3x = 3y \end{cases}$

9. $\begin{cases} y = -2x + 6 \\ 3y = x - 3 \end{cases}$

10. $\begin{cases} y = 5x - 1 \\ 10x = 2y + 2 \end{cases}$

11. $\begin{cases} 3x - 5y = 17 \\ 6x + 5y = 10 \end{cases}$

12. $\begin{cases} y = \frac{4}{3}x - 9 \\ y = x + 3 \end{cases}$

13. $\begin{cases} 4x - 7y = 11 \\ x + 2y = 10 \end{cases}$

14. $\begin{cases} 21x + 14y = 7 \\ 12x + 8y = 16 \end{cases}$

This page intentionally left blank

Lesson 29: Word Problems

Classwork

Example 1

The sum of two numbers is 361, and the difference between the two numbers is 173. What are the two numbers?

Example 2

There are 356 eighth-grade students at Euclid's Middle School. Thirty-four more than four times the number of girls is equal to half the number of boys. How many boys are in eighth grade at Euclid's Middle School? How many girls?

Example 3

A family member has some five-dollar bills and one-dollar bills in her wallet. Altogether she has 18 bills and a total of $62. How many of each bill does she have?

Example 4

A friend bought 2 boxes of pencils and 8 notebooks for school, and it cost him $11. He went back to the store the same day to buy school supplies for his younger brother. He spent $11.25 on 3 boxes of pencils and 5 notebooks. How much would 7 notebooks cost?

EUREKA
MATH

Exercises

1. A farm raises cows and chickens. The farmer has a total of 42 animals. One day he counts the legs of all of his animals and realizes he has a total of 114. How many cows does the farmer have? How many chickens?

2. The length of a rectangle is 4 times the width. The perimeter of the rectangle is 45 inches. What is the area of the rectangle?

3. The sum of the measures of angles x and y is 127°. If the measure of $\angle x$ is 34° more than half the measure of $\angle y$, what is the measure of each angle?

Lesson 29: Word Problems

EUREKA
MATH

Problem Set

1. Two numbers have a sum of 1,212 and a difference of 518. What are the two numbers?

2. The sum of the ages of two brothers is 46. The younger brother is 10 more than a third of the older brother's age. How old is the younger brother?

3. One angle measures 54 more degrees than 3 times another angle. The angles are supplementary. What are their measures?

4. Some friends went to the local movie theater and bought four large buckets of popcorn and six boxes of candy. The total for the snacks was $46.50. The last time you were at the theater, you bought a large bucket of popcorn and a box of candy, and the total was $9.75. How much would 2 large buckets of popcorn and 3 boxes of candy cost?

5. You have 59 total coins for a total of $12.05. You only have quarters and dimes. How many of each coin do you have?

6. A piece of string is 112 inches long. Isabel wants to cut it into 2 pieces so that one piece is three times as long as the other. How long is each piece?

This page intentionally left blank

Lesson 30: Conversion Between Celsius and Fahrenheit

Mathematical Modeling Exercise

(1) If t is a number, what is the degree in Fahrenheit that corresponds to $t°C$?

(2) If t is a number, what is the degree in Fahrenheit that corresponds to $(-t)°C$?

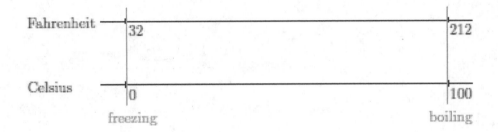

Exercises

Determine the corresponding Fahrenheit temperature for the given Celsius temperatures in Exercises 1–5.

1. How many degrees Fahrenheit is 25°C?

2. How many degrees Fahrenheit is 42°C?

3. How many degrees Fahrenheit is 94°C?

4. How many degrees Fahrenheit is 63°C?

5. How many degrees Fahrenheit is t°C?

EUREKA
MATH

Problem Set

1. Does the equation $t°C = (32 + 1.8t)°F$ work for any rational number t? Check that it does with $t = 8\frac{2}{3}$ and $t = -8\frac{2}{3}$.

2. Knowing that $t°C = \left(32 + \frac{9}{5}t\right)°F$ for any rational number t, show that for any rational number d,

$$d°F = \left(\frac{5}{9}(d - 32)\right)°C.$$

3. Drake was trying to write an equation to help him predict the cost of his monthly phone bill. He is charged $35 just for having a phone, and his only additional expense comes from the number of texts that he sends. He is charged $0.05 for each text. Help Drake out by completing parts (a)–(f).

   a. How much was his phone bill in July when he sent 750 texts?

   b. How much was his phone bill in August when he sent 823 texts?

   c. How much was his phone bill in September when he sent 579 texts?

   d. Let y represent the total cost of Drake's phone bill. Write an equation that represents the total cost of his phone bill in October if he sends t texts.

   e. Another phone plan charges $20 for having a phone and $0.10 per text. Let y represent the total cost of the phone bill for sending t texts. Write an equation to represent his total bill.

   f. Write your equations in parts (d) and (e) as a system of linear equations, and solve. Interpret the meaning of the solution in terms of the phone bill.

This page intentionally left blank

Lesson 31: System of Equations Leading to Pythagorean Triples

Classwork

Exercises

1. Identify two Pythagorean triples using the known triple 3, 4, 5 (other than 6, 8, 10).

2. Identify two Pythagorean triples using the known triple 5, 12, 13.

3. Identify two triples using either 3, 4, 5 or 5, 12, 13.

Use the system $\begin{cases} x + y = \dfrac{t}{s} \\ x - y = \dfrac{s}{t} \end{cases}$ to find Pythagorean triples for the given values of s and t. Recall that the solution in the form of $\left(\dfrac{c}{b}, \dfrac{a}{b}\right)$ is the triple a, b, c.

4. $s = 4, t = 5$

5. $s = 7, t = 10$

6. $s = 1, t = 4$

Lesson 31:   System of Equations Leading to Pythagorean Triples

EUREKA
MATH™

7. Use a calculator to verify that you found a Pythagorean triple in each of the Exercises 4–6. Show your work below.

Lesson Summary

A Pythagorean triple is a set of three positive integers that satisfies the equation $a^2 + b^2 = c^2$.

An infinite number of Pythagorean triples can be found by multiplying the numbers of a known triple by a whole number. For example, 3, 4, 5 is a Pythagorean triple. Multiply each number by 7, and then you have 21, 28, 35, which is also a Pythagorean triple.

The system of linear equations, $\begin{cases} x + y = \dfrac{t}{s} \\ x - y = \dfrac{s}{t} \end{cases}$, can be used to find Pythagorean triples, just like the Babylonians did 4,000 years ago.

Problem Set

1. Explain in terms of similar triangles why it is that when you multiply the known Pythagorean triple 3, 4, 5 by 12, it generates a Pythagorean triple.

2. Identify three Pythagorean triples using the known triple 8, 15, 17.

3. Identify three triples (numbers that satisfy $a^2 + b^2 = c^2$, but a, b, c are not whole numbers) using the triple 8, 15, 17.

Use the system $\begin{cases} x + y = \dfrac{t}{s} \\ x - y = \dfrac{s}{t} \end{cases}$ to find Pythagorean triples for the given values of s and t. Recall that the solution, in the form of $\left(\dfrac{c}{b}, \dfrac{a}{b}\right)$, is the triple a, b, c.

4. $s = 2, t = 9$

5. $s = 6, t = 7$

6. $s = 3, t = 4$

7. Use a calculator to verify that you found a Pythagorean triple in each of the Problems 4–6. Show your work.

EUREKA
MATH™

Eureka Math
Grade 8
Module 5

Special thanks go to the Gordon A. Cain Center and to the Department of Mathematics at Louisiana State University for their support in the development of *Eureka Math*.

For a free *Eureka Math* Teacher Resource Pack, Parent Tip Sheets, and more please visit www.Eureka.tools

Published by the non-profit Great Minds

Copyright © 2015 Great Minds. No part of this work may be reproduced, sold, or commercialized, in whole or in part, without written permission from Great Minds. Non-commercial use is licensed pursuant to a Creative Commons Attribution-NonCommercial-ShareAlike 4.0 license; for more information, go to http://greatminds.net/maps/math/copyright. "Great Minds" and "Eureka Math" are registered trademarks of Great Minds.

Printed in the U.S.A.
This book may be purchased from the publisher at eureka-math.org
1 2 3 4 5 6 7 8 BAB 25 24 23 22 21

ISBN 978-1-63255-321-8

Lesson 1: The Concept of a Function

Classwork

Example 1

Suppose a moving object travels 256 feet in 4 seconds. Assume that the object travels at a constant speed, that is, the motion of the object can be described by a linear equation. Write a linear equation in two variables to represent the situation, and use the equation to predict how far the object has moved at the four times shown.

Number of seconds in motion (x)	Distance traveled in feet (y)
1	
2	
3	
4	

Example 2

The object, a stone, is dropped from a height of 256 feet. It takes exactly 4 seconds for the stone to hit the ground. How far does the stone drop in the first 3 seconds? What about the last 3 seconds? Can we assume constant speed in this situation? That is, can this situation be expressed using a linear equation?

Number of seconds (x)	Distance traveled in feet (y)
1	
2	
3	
4	

Exercises 1–6

Use the table to answer Exercises 1–5.

Number of seconds (x)	Distance traveled in feet (y)
0.5	4
1	16
1.5	36
2	64
2.5	100
3	144
3.5	196
4	256

1. Name two predictions you can make from this table.

2. Name a prediction that would require more information.

3. What is the average speed of the object between 0 and 3 seconds? How does this compare to the average speed calculated over the same interval in Example 1?

$$\text{Average Speed} = \frac{\text{distance traveled over a given time interval}}{\text{time interval}}$$

EUREKA
MATH™

4. Take a closer look at the data for the falling stone by answering the questions below.

   a. How many feet did the stone drop between 0 and 1 second?

   b. How many feet did the stone drop between 1 and 2 seconds?

   c. How many feet did the stone drop between 2 and 3 seconds?

   d. How many feet did the stone drop between 3 and 4 seconds?

   e. Compare the distances the stone dropped from one time interval to the next. What do you notice?

5. What is the average speed of the stone in each interval 0.5 second? For example, the average speed over the interval from 3.5 seconds to 4 seconds is

$$\frac{\text{distance traveled over a given time interval}}{\text{time interval}} = \frac{256 - 196}{4 - 3.5} = \frac{60}{0.5} = 120; 120 \text{ feet per second}$$

Repeat this process for every half-second interval. Then answer the question that follows.

   a. Interval between 0 and 0.5 second: b. Interval between 0.5 and 1 second:

   c. Interval between 1 and 1.5 seconds: d. Interval between 1.5 and 2 seconds:

e.   Interval between 2 and 2.5 seconds: f.   Interval between 2.5 and 3 seconds:

g.   Interval between 3 and 3.5 seconds:

h.   Compare the average speed between each time interval. What do you notice?

6.   Is there any pattern to the data of the falling stone? Record your thoughts below.

Time of interval in seconds (t)	1	2	3	4
Distance stone fell in feet (y)	16	64	144	256

EUREKA
MATH

Lesson Summary

A *function* is a rule that assigns to each value of one quantity a single value of a second quantity. Even though we might not have a formula for that rule, we see that functions do arise in real-life situations

---

Problem Set

A ball is thrown across the field from point A to point B. It hits the ground at point B. The path of the ball is shown in the diagram below. The x-axis shows the horizontal distance the ball travels in feet, and the y-axis shows the height of the ball in feet. Use the diagram to complete parts (a)–(f).

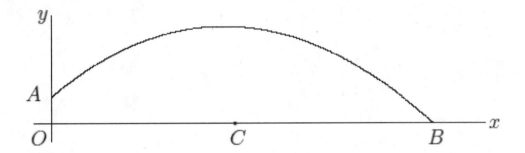

a.   Suppose point A is approximately 6 feet above ground and that at time $t = 0$ the ball is at point A. Suppose the length of OB is approximately 88 feet. Include this information on the diagram.

b.   Suppose that after 1 second, the ball is at its highest point of 22 feet (above point C) and has traveled a horizontal distance of 44 feet. What are the approximate coordinates of the ball at the following values of t: 0.25, 0.5, 0.75, 1, 1.25, 1.5, 1.75, and 2.

c.   Use your answer from part (b) to write two predictions.

d.   What is happening to the ball when it has coordinates $(88, 0)$?

e.   Why do you think the ball is at point $(0, 6)$ when $t = 0$? In other words, why isn't the height of the ball 0?

f.   Does the graph allow us to make predictions about the height of the ball at all points?

This page intentionally left blank

Lesson 2: Formal Definition of a Function

Classwork

Exercises 1–5

1. Let D be the distance traveled in time t. Use the equation $D = 16t^2$ to calculate the distance the stone dropped for the given time t.

Time in seconds	0.5	1	1.5	2	2.5	3	3.5	4
Distance stone fell in feet by that time								

a. Are the distances you calculated equal to the table from Lesson 1?

b. Does the function $D = 16t^2$ accurately represent the distance the stone fell after a given time t? In other words, does the function described by this rule assign to t the correct distance? Explain.

2. Can the table shown below represent values of a function? Explain.

Input (x)	1	3	5	5	9
Output (y)	7	16	19	20	28

3. Can the table shown below represent values of a function? Explain.

Input (x)	0.5	7	7	12	15
Output (y)	1	15	10	23	30

4. Can the table shown below represent values of a function? Explain.

Input (x)	10	20	50	75	90
Output (y)	32	32	156	240	288

5. It takes Josephine 34 minutes to complete her homework assignment of 10 problems. If we assume that she works at a constant rate, we can describe the situation using a function.

a. Predict how many problems Josephine can complete in 25 minutes.

Lesson 2:   Formal Definition of a Function

EUREKA
MATH™

b. Write the two-variable linear equation that represents Josephine's constant rate of work.

c. Use the equation you wrote in part (b) as the formula for the function to complete the table below. Round your answers to the hundredths place.

Time taken to complete problems (x)	5	10	15	20	25
Number of problems completed (y)	1.47				

After 5 minutes, Josephine was able to complete 1.47 problems, which means that she was able to complete 1 problem, then get about halfway through the next problem.

d. Compare your prediction from part (a) to the number you found in the table above.

e. Use the formula from part (b) to compute the number of problems completed when $x = -7$. Does your answer make sense? Explain.

f. For this problem, we assumed that Josephine worked at a constant rate. Do you think that is a reasonable assumption for this situation? Explain.

EUREKA
MATH™

Lesson 2:   Formal Definition of a Function

© 2015 Great Minds. eureka-math.org
G8-M3M4M5-SE-B2-1.3.1-01.2016

S.9

Lesson Summary

A *function* is a correspondence between a set (whose elements are called *inputs*) and another set (whose elements are called *outputs*) such that each input corresponds to one and only one output.

Sometimes the phrase *exactly one output* is used instead of *one and only one output* in the definition of function (they mean the same thing). Either way, it is this fact, that there is one and only one output for each input, which makes functions predictive when modeling real life situations.

Furthermore, the correspondence in a function is often given by a *rule* (or *formula*). For example, the output is equal to the number found by substituting an input number into the variable of a one-variable expression and evaluating.

Functions are sometimes described as an *input–output machine*. For example, given a function D, the input is time t, and the output is the distance traveled in t seconds.

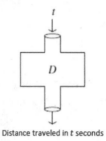

Distance traveled in t seconds

Problem Set

1. The table below represents the number of minutes Francisco spends at the gym each day for a week. Does the data shown below represent values of a function? Explain.

Day (x)	1	2	3	4	5	6	7
Time in minutes (y)	35	45	30	45	35	0	0

2. Can the table shown below represent values of a function? Explain.

Input (x)	9	8	7	8	9
Output (y)	11	15	19	24	28

EUREKA
MATH™

3. Olivia examined the table of values shown below and stated that a possible rule to describe this function could be $y = -2x + 9$. Is she correct? Explain.

Input (x)	−4	0	4	8	12	16	20	24
Output (y)	17	9	1	−7	−15	−23	−31	−39

4. Peter said that the set of data in part (a) describes a function, but the set of data in part (b) does not. Do you agree? Explain why or why not.

a.

Input (x)	1	2	3	4	5	6	7	8
Output (y)	8	10	32	6	10	27	156	4

b.

Input (x)	−6	−15	−9	−3	−2	−3	8	9
Output (y)	0	−6	8	14	1	2	11	41

5. A function can be described by the rule $y = x^2 + 4$. Determine the corresponding output for each given input.

Input (x)	−3	−2	−1	0	1	2	3	4
Output (y)								

6. Examine the data in the table below. The inputs and outputs represent a situation where constant rate can be assumed. Determine the rule that describes the function.

Input (x)	−1	0	1	2	3	4	5	6
Output (y)	3	8	13	18	23	28	33	38

7. Examine the data in the table below. The inputs represent the number of bags of candy purchased, and the outputs represent the cost. Determine the cost of one bag of candy, assuming the price per bag is the same no matter how much candy is purchased. Then, complete the table.

Bags of candy (x)	1	2	3	4	5	6	7	8
Cost in Dollars (y)				5.00	6.25			10.00

a. Write the rule that describes the function.

b. Can you determine the value of the output for an input of $x = -4$? If so, what is it?

c. Does an input of -4 make sense in this situation? Explain.

8. Each and every day a local grocery store sells 2 pounds of bananas for $1.00. Can the cost of 2 pounds of bananas be represented as a function of the day of the week? Explain.

9. Write a brief explanation to a classmate who was absent today about why the table in part (a) is a function and the table in part (b) is not.

a.

Input (x)	−1	−2	−3	−4	4	3	2	1
Output (y)	81	100	320	400	400	320	100	81

b.

Input (x)	1	6	−9	−2	1	−10	8	14
Output (y)	2	6	−47	−8	19	−2	15	31

Lesson 3: Linear Functions and Proportionality

Classwork

Example 1

In the last lesson, we looked at several tables of values showing the inputs and outputs of functions. For instance, one table showed the costs of purchasing different numbers of bags of candy:

Bags of candy (x)	1	2	3	4	5	6	7	8
Cost in Dollars (y)	1.25	2.50	3.75	5.00	6.25	7.50	8.75	10.00

Example 2

Walter walks at a constant speed of 8 miles every 2 hours. Describe a linear function for the number of miles he walks in x hours. What is a reasonable range of x-values for this function?

Example 3

Veronica runs at a constant speed. The distance she runs is a function of the time she spends running. The function has the table of values shown below.

Time in minutes (x)	8	16	24	32
Distance run in miles (y)	1	2	3	4

Example 4

Water flows from a faucet into a bathtub at the constant rate of 7 gallons of water pouring out every 2 minutes. The bathtub is initially empty, and its plug is in. Determine the rule that describes the volume of water in the tub as a function of time. If the tub can hold 50 gallons of water, how long will it take to fill the tub?

Now assume that you are filling the same 50-gallon bathtub with water flowing in at the constant rate of 3.5 gallons per minute, but there were initially 8 gallons of water in the tub. Will it still take about 14 minutes to fill the tub?

Time in minutes (x)	0	3	6	9	12
Total volume in tub in gallons (y)					

Lesson 3:   Linear Functions and Proportionality

EUREKA MATH

© 2015 Great Minds. eureka-math.org
G8-M3M4M5-SE-B2-1.3.1-01.2016

Example 5

Water flows from a faucet at a constant rate. Assume that 6 gallons of water are already in a tub by the time we notice the faucet is on. This information is recorded in the first column of the table below. The other columns show how many gallons of water are in the tub at different numbers of minutes since we noticed the running faucet.

Time in minutes (x)	0	3	5	9
Total volume in tub in gallons (y)	6	9.6	12	16.8

Exercises 1–3

1. Hana claims she mows lawns at a constant rate. The table below shows the area of lawn she can mow over different time periods.

Number of minutes (x)	5	20	30	50
Area mowed in square feet (y)	36	144	216	360

   a. Is the data presented consistent with the claim that the area mowed is a linear function of time?

b.   Describe in words the function in terms of area mowed and time.

c.   At what rate does Hana mow lawns over a 5-minute period?

d.   At what rate does Hana mow lawns over a 20-minute period?

e.   At what rate does Hana mow lawns over a 30-minute period?

f.   At what rate does Hana mow lawns over a 50-minute period?

g.   Write the equation that describes the area mowed, y, in square feet, as a linear function of time, x, in minutes.

h.   Describe any limitations on the possible values of x and y.

EUREKA
MATH™

    i.   What number does the function assign to $x = 24$? That is, what area of lawn can be mowed in 24 minutes?

    j.   According to this work, how many minutes would it take to mow an area of 400 square feet?

2.   A linear function has the table of values below. The information in the table shows the total volume of water, in gallons, that flows from a hose as a function of time, the number of minutes the hose has been running.

Time in minutes (x)	10	25	50	70
Total volume of water in gallons (y)	44	110	220	308

    a.   Describe the function in terms of volume and time.

    b.   Write the rule for the volume of water in gallons, y, as a linear function of time, x, given in minutes.

    c.   What number does the function assign to 250? That is, how many gallons of water flow from the hose during a period of 250 minutes?

d. The average swimming pool holds about 17,300 gallons of water. Suppose such a pool has already been filled one quarter of its volume. Write an equation that describes the volume of water in the pool if, at time 0 minutes, we use the hose described above to start filling the pool.

e. Approximately how many hours will it take to finish filling the pool?

3. Recall that a linear function can be described by a rule in the form of $y = mx + b$, where m and b are constants. A particular linear function has the table of values below.

Input (x)	0	4	10	11	15	20	23
Output (y)	4	24	54	59			

a. What is the equation that describes the function?

b. Complete the table using the rule.

EUREKA MATH

> **Lesson Summary**
>
> A linear equation $y = mx + b$ describes a rule for a function. We call any function defined by a linear equation a linear function.
>
> Problems involving a constant rate of change or a proportional relationship can be described by linear functions.

Problem Set

1. A food bank distributes cans of vegetables every Saturday. The following table shows the total number of cans they have distributed since the beginning of the year. Assume that this total is a linear function of the number of weeks that have passed.

Number of weeks (x)	1	12	20	45
Total number of cans of vegetables distributed (y)	180	2,160	3,600	8,100

   a. Describe the function being considered in words.

   b. Write the linear equation that describes the total number of cans handed out, y, in terms of the number of weeks, x, that have passed.

   c. Assume that the food bank wants to distribute 20,000 cans of vegetables. How long will it take them to meet that goal?

   d. The manager had forgotten to record that they had distributed 35,000 cans on January 1. Write an adjusted linear equation to reflect this forgotten information.

   e. Using your function in part (d), determine how long in years it will take the food bank to hand out 80,000 cans of vegetables.

2. A linear function has the table of values below. It gives the number of miles a plane travels over a given number of hours while flying at a constant speed.

Number of hours traveled (x)	2.5	4	4.2
Distance in miles (y)	1,062.5	1,700	1,785

   a. Describe in words the function given in this problem.

   b. Write the equation that gives the distance traveled, y, in miles, as a linear function of the number of hours, x, spent flying.

   c. Assume that the airplane is making a trip from New York to Los Angeles, which is a journey of approximately 2,475 miles. How long will it take the airplane to get to Los Angeles?

   d. If the airplane flies for 8 hours, how many miles will it cover?

3. A linear function has the table of values below. It gives the number of miles a car travels over a given number of hours.

Number of hours traveled (x)	3.5	3.75	4	4.25
Distance in miles (y)	203	217.5	232	246.5

a. Describe in words the function given.

b. Write the equation that gives the distance traveled, in miles, as a linear function of the number of hours spent driving.

c. Assume that the person driving the car is going on a road trip to reach a location 500 miles from her starting point. How long will it take the person to get to the destination?

4. A particular linear function has the table of values below.

Input (x)	2	3	8	11	15	20	23
Output (y)	7	10		34		61	

a. What is the equation that describes the function?

b. Complete the table using the rule.

5. A particular linear function has the table of values below.

Input (x)	0	5	8	13	15	18	21
Output (y)	6	11	14		21		

a. What is the rule that describes the function?

b. Complete the table using the rule.

Lesson 4: More Examples of Functions

Classwork

Example 1

Classify each of the functions described below as either discrete or not discrete.

a) The function that assigns to each whole number the cost of buying that many cans of beans in a particular grocery store.

b) The function that assigns to each time of day one Wednesday the temperature of Sammy's fever at that time.

c) The function that assigns to each real number its first digit.

d) The function that assigns to each day in the year 2015 my height at noon that day.

e) The function that assigns to each moment in the year 2015 my height at that moment.

f) The function that assigns to each color the first letter of the name of that color.

g) The function that assigns the number 23 to each and every real number between 20 and 30.6.

h) The function that assigns the word YES to every yes/no question.

i) The function that assigns to each height directly above the North Pole the temperature of the air at that height right at this very moment.

Example 2

Water flows from a faucet into a bathtub at a constant rate of 7 gallons of water every 2 minutes. Regard the volume of water accumulated in the tub as a function of the number of minutes the faucet has been on. Is this function discrete or not discrete?

Example 3

You have just been served freshly made soup that is so hot that it cannot be eaten. You measure the temperature of the soup, and it is 210°F. Since 212°F is boiling, there is no way it can safely be eaten yet. One minute after receiving the soup, the temperature has dropped to 203°F. If you assume that the rate at which the soup cools is constant, write an equation that would describe the temperature of the soup over time.

Example 4

Consider the function that assigns to each of nine baseball players, numbered 1 through 9, his height. The data for this function is given below. Call the function G.

Player Number	Height
1	5'11"
2	5'4"
3	5'9"
4	5'6"
5	6'3"
6	6'8"
7	5'9"
8	5'10"
9	6'2"

EUREKA
MATH

Exercises 1–3

1. At a certain school, each bus in its fleet of buses can transport 35 students. Let B be the function that assigns to each count of students the number of buses needed to transport that many students on a field trip.

   When Jinpyo thought about the situation, he drew the following table of values and wrote the formula $B = \frac{x}{35}$. Here x is the count of students, and B is the number of buses needed to transport that many students. He concluded that B is a linear function.

Number of students (x)	35	70	105	140
Number of buses (B)	1	2	3	4

   Alicia looked at Jinpyo's work and saw no errors with his arithmetic. But she said that the function is not actually linear.

   a. Alicia is right. Explain why B is not a linear function.

   b. Is B a discrete function?

2. A linear function has the table of values below. It gives the costs of purchasing certain numbers of movie tickets.

Number of tickets (x)	3	6	9	12
Total cost in dollars (y)	27.75	55.50	83.25	111.00

a. Write the linear function that represents the total cost, y, for x tickets purchased.

b. Is the function discrete? Explain.

c. What number does the function assign to 4? What do the question and your answer mean?

3. A function produces the following table of values.

Input	Output
Banana	B
Cat	C
Flippant	F
Oops	O
Slushy	S

a. Make a guess as to the rule this function follows. Each input is a word from the English language.

b. Is this function discrete?

EUREKA
MATH™

Lesson Summary

Functions are classified as either discrete or not discrete.

Discrete functions admit only individually separate input values (such as whole numbers of students, or words of the English language). Functions that are not discrete admit any input value within a range of values (fractional values, for example).

Functions that describe motion or smooth changes over time, for example, are typically not discrete.

Problem Set

1. The costs of purchasing certain volumes of gasoline are shown below. We can assume that there is a linear relationship between x, the number of gallons purchased, and y, the cost of purchasing that many gallons.

Number of gallons (x)	5.4	6	15	17
Total cost in dollars (y)	19.71	21.90	54.75	62.05

   a. Write an equation that describes y as a linear function of x.

   b. Are there any restrictions on the values x and y can adopt?

   c. Is the function discrete?

   d. What number does the linear function assign to 20? Explain what your answer means.

2. A function has the table of values below. Examine the information in the table to answer the questions below.

Input	Output
one	3
two	3
three	5
four	4
five	4
six	3
seven	5

   a. Describe the function.

   b. What number would the function assign to the word *eleven*?

3. The table shows the distances covered over certain counts of hours traveled by a driver driving a car at a constant speed.

Number of hours driven (x)	3	4	5	6
Total miles driven (y)	141	188	235	282

a. Write an equation that describes y, the number of miles covered, as a linear function of x, number of hours driven.

b. Are there any restrictions on the value x and y can adopt?

c. Is the function discrete?

d. What number does the function assign to 8? Explain what your answer means.

e. Use the function to determine how much time it would take to drive 500 miles.

4. Consider the function that assigns to each time of a particular day the air temperature at a specific location in Ithaca, NY. The following table shows the values of this function at some specific times.

12:00 noon	92°F
1:00 p.m.	90.5°F
2:00 p.m.	89°F
4:00 p.m.	86°F
8:00 p.m.	80°F

a. Let y represent the air temperature at time x hours past noon. Verify that the data in the table satisfies the linear equation $y = 92 - 1.5x$.

b. Are there any restrictions on the types of values x and y can adopt?

c. Is the function discrete?

d. According to the linear function of part (a), what will the air temperature be at 5:30 p.m.?

e. Is it reasonable to assume that this linear function could be used to predict the temperature for 10:00 a.m. the following day or a temperature at any time on a day next week? Give specific examples in your explanation.

EUREKA
MATH™

Lesson 5: Graphs of Functions and Equations

Classwork

Exploratory Challenge/Exercises 1–3

1. The distance that Giselle can run is a function of the amount of time she spends running. Giselle runs 3 miles in 21 minutes. Assume she runs at a constant rate.

   a. Write an equation in two variables that represents her distance run, y, as a function of the time, x, she spends running.

   b. Use the equation you wrote in part (a) to determine how many miles Giselle can run in 14 minutes.

   c. Use the equation you wrote in part (a) to determine how many miles Giselle can run in 28 minutes.

   d. Use the equation you wrote in part (a) to determine how many miles Giselle can run in 7 minutes.

e. For a given input x of the function, a time, the matching output of the function, y, is the distance Giselle ran in that time. Write the inputs and outputs from parts (b)–(d) as ordered pairs, and plot them as points on a coordinate plane.

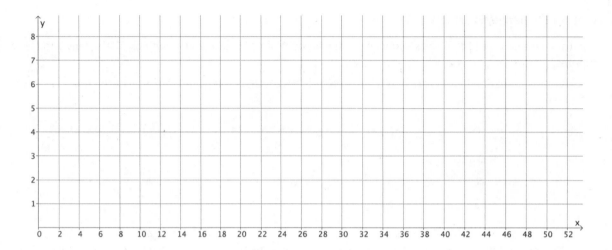

f. What do you notice about the points you plotted?

g. Is the function discrete?

h. Use the equation you wrote in part (a) to determine how many miles Giselle can run in 36 minutes. Write your answer as an ordered pair, as you did in part (e), and include the point on the graph. Is the point in a place where you expected it to be? Explain.

EUREKA
MATH™

i.   Assume you used the rule that describes the function to determine how many miles Giselle can run for any given time and wrote each answer as an ordered pair. Where do you think these points would appear on the graph?

j.   What do you think the graph of all the input/output pairs would look like? Explain.

k.   Connect the points you have graphed to make a line. Select a point on the graph that has integer coordinates. Verify that this point has an output that the function would assign to the input.

l.   Sketch the graph of the equation $y = \frac{1}{7}x$ using the same coordinate plane in part (e). What do you notice about the graph of all the input/output pairs that describes Giselle's constant rate of running and the graph of the equation $y = \frac{1}{7}x$?

2. Sketch the graph of the equation $y = x^2$ for positive values of x. Organize your work using the table below, and then answer the questions that follow.

x	y
0	
1	
2	
3	
4	
5	
6	

a. Plot the ordered pairs on the coordinate plane.

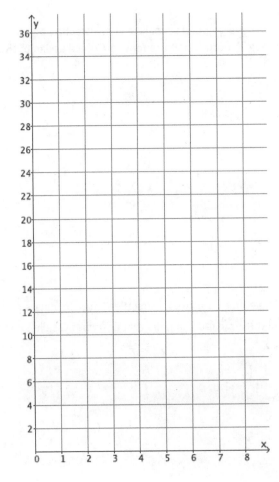

Lesson 5:   Graphs of Functions and Equations

EUREKA
MATH™

b.   What shape does the graph of the points appear to take?

c.   Is this equation a linear equation? Explain.

d.   Consider the function that assigns to each square of side length s units its area A square units. Write an equation that describes this function.

e.   What do you think the graph of all the input/output pairs (s, A) of this function will look like? Explain.

f.   Use the function you wrote in part (d) to determine the area of a square with side length 2.5 units. Write the input and output as an ordered pair. Does this point appear to belong to the graph of $y = x^2$?

3. The number of devices a particular manufacturing company can produce is a function of the number of hours spent making the devices. On average, 4 devices are produced each hour. Assume that devices are produced at a constant rate.

    a. Write an equation in two variables that describes the number of devices, y, as a function of the time the company spends making the devices, x.

    b. Use the equation you wrote in part (a) to determine how many devices are produced in 8 hours.

    c. Use the equation you wrote in part (a) to determine how many devices are produced in 6 hours.

    d. Use the equation you wrote in part (a) to determine how many devices are produced in 4 hours.

EUREKA
MATH™

e.   The input of the function, x, is time, and the output of the function, y, is the number of devices produced. Write the inputs and outputs from parts (b)–(d) as ordered pairs, and plot them as points on a coordinate plane.

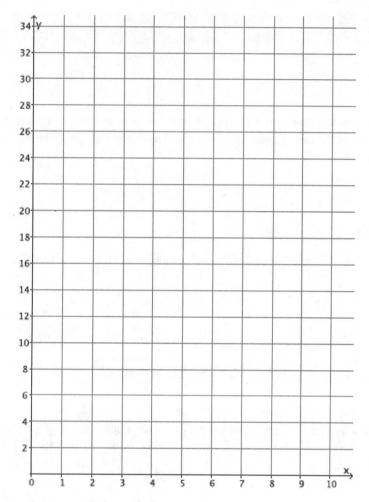

f.   What shape does the graph of the points appear to take?

g.   Is the function discrete?

EUREKA
MATH™

Lesson 5:   Graphs of Functions and Equations

S.33

© 2015 Great Minds. eureka-math.org
G8-M3M4M5-SE-B2-1.3.1-01.2016

h. Use the equation you wrote in part (a) to determine how many devices are produced in 1.5 hours. Write your answer as an ordered pair, as you did in part (e), and include the point on the graph. Is the point in a place where you expected it to be? Explain.

i. Assume you used the equation that describes the function to determine how many devices are produced for any given time and wrote each answer as an ordered pair. Where do you think these points would appear on the graph?

j. What do you think the graph of all possible input/output pairs will look like? Explain.

k. Connect the points you have graphed to make a line. Select a point on the graph that has integer coordinates. Verify that this point has an output that the function would assign to the input.

l. Sketch the graph of the equation $y = 4x$ using the same coordinate plane in part (e). What do you notice about the graph of the input/output pairs that describes the company's constant rate of producing devices and the graph of the equation $y = 4x$?

Exploratory Challenge/Exercise 4

4. Examine the three graphs below. Which, if any, could represent the graph of a function? Explain why or why not for each graph.

Graph 1:

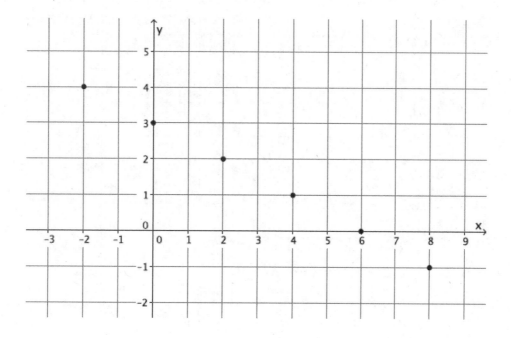

Graph 2:

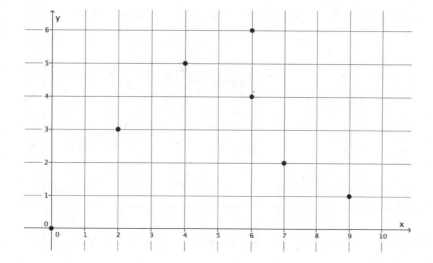

Graph 3:

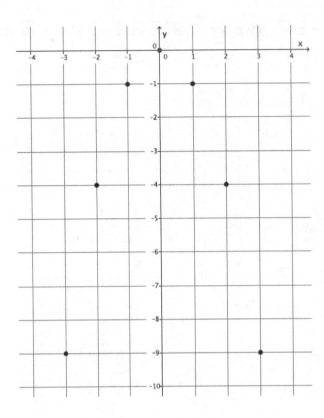

EUREKA
MATH

Lesson Summary

The graph of a function is defined to be the set of all points (x, y) with x an input for the function and y its matching output.

If a function can be described by an equation, then the graph of the function is the same as the graph of the equation that represents it (at least at points which correspond to valid inputs of the function).

It is not possible for two different points in the plot of the graph of a function to have the same x-coordinate.

Problem Set

1. The distance that Scott walks is a function of the time he spends walking. Scott can walk $\frac{1}{2}$ mile every 8 minutes. Assume he walks at a constant rate.

   a. Predict the shape of the graph of the function. Explain.

   b. Write an equation to represent the distance that Scott can walk in miles, y, in x minutes.

   c. Use the equation you wrote in part (b) to determine how many miles Scott can walk in 24 minutes.

   d. Use the equation you wrote in part (b) to determine how many miles Scott can walk in 12 minutes.

   e. Use the equation you wrote in part (b) to determine how many miles Scott can walk in 16 minutes.

   f. Write your inputs and corresponding outputs as ordered pairs, and then plot them on a coordinate plane.

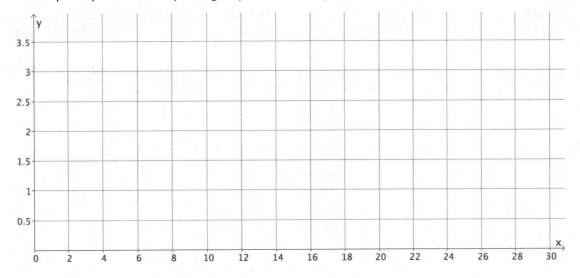

   g. What shape does the graph of the points appear to take? Does it match your prediction?

   h. Connect the points to make a line. What is the equation of the line?

2. Graph the equation $y = x^3$ for positive values of x. Organize your work using the table below, and then answer the questions that follow.

x	y
0	
0.5	
1	
1.5	
2	
2.5	

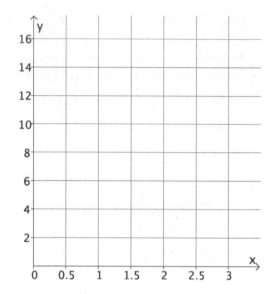

a. Plot the ordered pairs on the coordinate plane.

b. What shape does the graph of the points appear to take?

c. Is this the graph of a linear function? Explain.

d. Consider the function that assigns to each positive real number s the volume V of a cube with side length s units. An equation that describes this function is $V = s^3$. What do you think the graph of this function will look like? Explain.

e. Use the function in part (d) to determine the volume of a cube with side length of 3 units. Write the input and output as an ordered pair. Does this point appear to belong to the graph of $y = x^3$?

3. Sketch the graph of the equation $y = 180(x - 2)$ for whole numbers. Organize your work using the table below, and then answer the questions that follow.

x	y
3	
4	
5	
6	

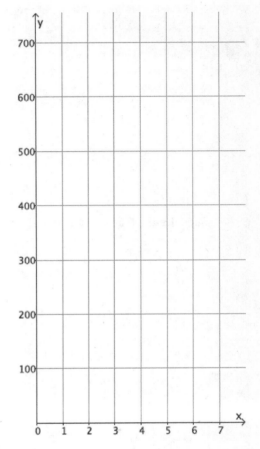

a. Plot the ordered pairs on the coordinate plane.

b. What shape does the graph of the points appear to take?

c. Is this graph a graph of a function? How do you know?

d. Is this a linear equation? Explain.

e. The sum S of interior angles, in degrees, of a polygon with n sides is given by $S = 180(n - 2)$. If we take this equation as defining S as a function of n, how do think the graph of this S will appear? Explain.

f. Is this function discrete? Explain.

4.   Examine the graph below. Could the graph represent the graph of a function? Explain why or why not.

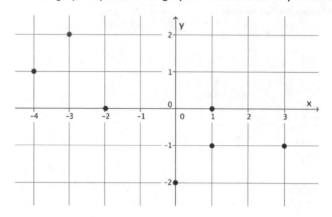

5.   Examine the graph below. Could the graph represent the graph of a function? Explain why or why not.

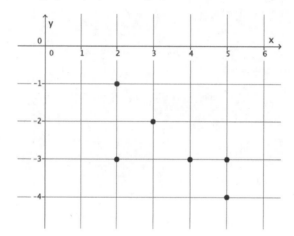

6.   Examine the graph below. Could the graph represent the graph of a function? Explain why or why not.

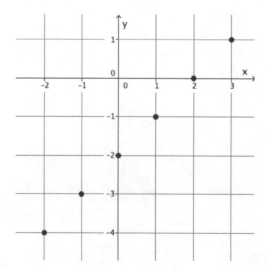

EUREKA
MATH™

Lesson 6: Graphs of Linear Functions and Rate of Change

Opening Exercise

A function is said to be linear if the rule defining the function can be described by a linear equation.

Functions 1, 2, and 3 have table-values as shown. Which of these functions appear to be linear? Justify your answers.

Input	Output
2	5
4	7
5	8
8	11

Input	Output
2	4
3	9
4	16
5	25

Input	Output
0	−3
1	1
2	6
3	9

© 2015 Great Minds. eureka-math.org
G8-M3M4M5-SE-B2-1.3.1-01.2016

Exercise

A function assigns the inputs shown the corresponding outputs given in the table below.

Input	Output
1	2
2	−1
4	−7
6	−13

a. Do you suspect the function is linear? Compute the rate of change of this data for at least three pairs of inputs and their corresponding outputs.

b. What equation seems to describe the function?

c. As you did not verify that the rate of change is constant across <u>all</u> input/output pairs, check that the equation you found in part (a) does indeed produce the correct output for each of the four inputs 1, 2, 4, and 6.

d. What will the graph of the function look like? Explain.

Lesson 6:   Graphs of Linear Functions and Rate of Change

Lesson Summary

If the rate of change for pairs of inputs and corresponding outputs for a function is the same for all pairs (constant), then the function is a linear function. It can thus be described by a linear equation $y = mx + b$.

The graph of a linear function will be a set of points contained in a line. If the linear function is discrete, then its graph will be a set of distinct collinear points. If the linear function is not discrete, then its graph will be a full straight line or a portion of the line (as appropriate for the context of the problem).

Problem Set

1. A function assigns to the inputs given the corresponding outputs shown in the table below.

Input	Output
3	9
9	17
12	21
15	25

    a. Does the function appear to be linear? Check at least three pairs of inputs and their corresponding outputs.

    b. Find a linear equation that describes the function.

    c. What will the graph of the function look like? Explain.

2. A function assigns to the inputs given the corresponding outputs shown in the table below.

Input	Output
−1	2
0	0
1	2
2	8
3	18

    a. Is the function a linear function?

    b. What equation describes the function?

3. A function assigns the inputs and corresponding outputs shown in the table below.

Input	Output
0.2	2
0.6	6
1.5	15
2.1	21

    a. Does the function appear to be linear? Check at least three pairs of inputs and their corresponding outputs.

    b. Find a linear equation that describes the function.

    c. What will the graph of the function look like? Explain.

4. Martin says that you only need to check the first and last input and output values to determine if the function is linear. Is he correct? Explain.

5. Is the following graph a graph of a linear function? How would you determine if it is a linear function?

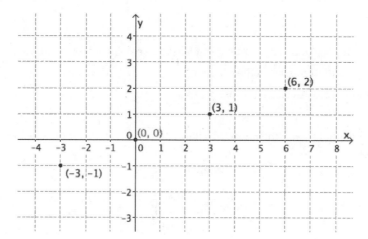

6. A function assigns to the inputs given the corresponding outputs shown in the table below.

Input	Output
−6	−6
−5	−5
−4	−4
−2	−2

    a. Does the function appear to be a linear function?

    b. What equation describes the function?

    c. What will the graph of the function look like? Explain.

Lesson 7: Comparing Linear Functions and Graphs

Classwork

Exploratory Challenge/Exercises 1–4

Each of Exercises 1–4 provides information about two functions. Use that information given to help you compare the two functions and answer the questions about them.

1. Alan and Margot each drive from City A to City B, a distance of 147 miles. They take the same route and drive at constant speeds. Alan begins driving at 1:40 p.m. and arrives at City B at 4:15 p.m. Margot's trip from City A to City B can be described with the equation $y = 64x$, where y is the distance traveled in miles and x is the time in minutes spent traveling. Who gets from City A to City B faster?

2. You have recently begun researching phone billing plans. Phone Company A charges a flat rate of $75 a month. A flat rate means that your bill will be $75 each month with no additional costs. The billing plan for Phone Company B is a linear function of the number of texts that you send that month. That is, the total cost of the bill changes each month depending on how many texts you send. The table below represents some inputs and the corresponding outputs that the function assigns.

Input (number of texts)	Output (cost of bill in dollars)
50	50
150	60
200	65
500	95

At what number of texts would the bill from each phone plan be the same? At what number of texts is Phone Company A the better choice? At what number of texts is Phone Company B the better choice?

© 2015 Great Minds. eureka-math.org
G8-M3M4M5-SE-B2-1.3.1-01.2016

3. The function that gives the volume of water, y, that flows from Faucet A in gallons during x minutes is a linear function with the graph shown. Faucet B's water flow can be described by the equation $y = \frac{5}{6}x$, where y is the volume of water in gallons that flows from the faucet during x minutes. Assume the flow of water from each faucet is constant. Which faucet has a faster rate of flow of water? Each faucet is being used to fill a tub with a volume of 50 gallons. How long will it take each faucet to fill its tub? How do you know?

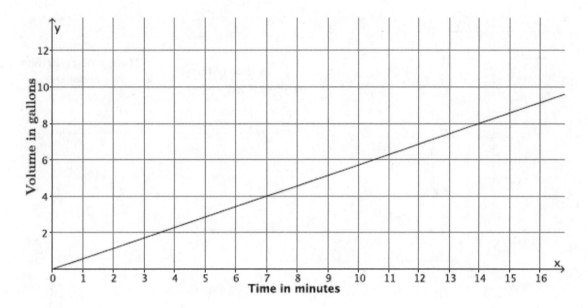

Suppose the tub being filled by Faucet A already had 15 gallons of water in it, and the tub being filled by Faucet B started empty. If now both faucets are turned on at the same time, which faucet will fill its tub fastest?

4. Two people, Adam and Bianca, are competing to see who can save the most money in one month. Use the table and the graph below to determine who will save the most money at the end of the month. State how much money each person had at the start of the competition. (Assume each is following a linear function in his or her saving habit.)

Adam's Savings:

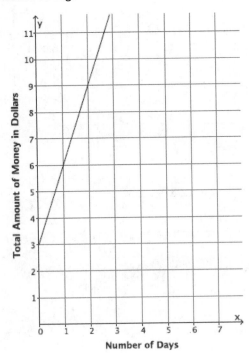

Number of Days

Bianca's Savings:

Input (Number of Days)	Output (Total amount of money in dollars)
5	17
8	26
12	38
20	62

Problem Set

1. The graph below represents the distance in miles, y, Car A travels in x minutes. The table represents the distance in miles, y, Car B travels in x minutes. It is moving at a constant rate. Which car is traveling at a greater speed? How do you know?

Car A:

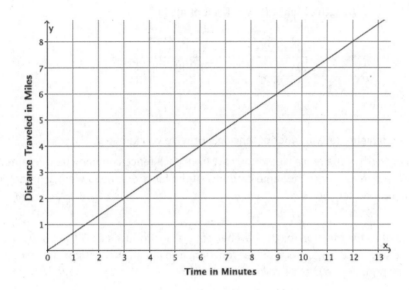

Car B:

Time in minutes (x)	Distance in miles (y)
15	12.5
30	25
45	37.5

2. The local park needs to replace an existing fence that is 6 feet high. Fence Company A charges $7,000 for building materials and $200 per foot for the length of the fence. Fence Company B charges are based solely on the length of the fence. That is, the total cost of the six-foot high fence will depend on how long the fence is. The table below represents some inputs and their corresponding outputs that the cost function for Fence Company B assigns. It is a linear function.

Input (length of fence in feet)	Output (cost of bill in dollars)
100	26,000
120	31,200
180	46,800
250	65,000

a. Which company charges a higher rate per foot of fencing? How do you know?

b. At what number of the length of the fence would the cost from each fence company be the same? What will the cost be when the companies charge the same amount? If the fence you need were 190 feet in length, which company would be a better choice?

3. The equation $y = 123x$ describes the function for the number of toys, y, produced at Toys Plus in x minutes of production time. Another company, #1 Toys, has a similar function, also linear, that assigns the values shown in the table below. Which company produces toys at a slower rate? Explain.

Time in minutes (x)	Toys Produced (y)
5	600
11	1,320
13	1,560

4. A train is traveling from City A to City B, a distance of 320 miles. The graph below shows the number of miles, y, the train travels as a function of the number of hours, x, that have passed on its journey. The train travels at a constant speed for the first four hours of its journey and then slows down to a constant speed of 48 miles per hour for the remainder of its journey.

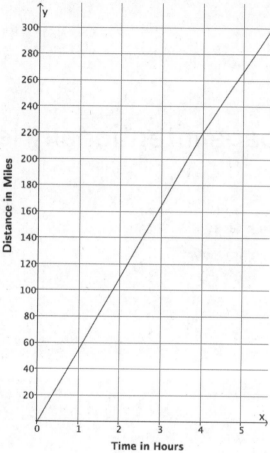

a. How long will it take the train to reach its destination?

b. If the train had not slowed down after 4 hours, how long would it have taken to reach its destination?

c. Suppose after 4 hours, the train increased its constant speed. How fast would the train have to travel to complete the destination in 1.5 hours?

5.

a. A hose is used to fill up a 1,200 gallon water truck. Water flows from the hose at a constant rate. After 10 minutes, there are 65 gallons of water in the truck. After 15 minutes, there are 82 gallons of water in the truck. How long will it take to fill up the water truck? Was the tank initially empty?

b. The driver of the truck realizes that something is wrong with the hose he is using. After 30 minutes, he shuts off the hose and tries a different hose. The second hose flows at a constant rate of 18 gallons per minute. How long now does it take to fill up the truck?

This page intentionally left blank

Lesson 8: Graphs of Simple Nonlinear Functions

Classwork

Exploratory Challenge/Exercises 1–3

1. Consider the function that assigns to each number x the value x^2.

    a. Do you think the function is linear or nonlinear? Explain.

    b. Develop a list of inputs and outputs for this function. Organize your work using the table below. Then, answer the questions that follow.

Input (x)	Output (x^2)
−5	
−4	
−3	
−2	
−1	
0	
1	
2	
3	
4	
5	

c.   Plot the inputs and outputs as ordered pairs defining points on the coordinate plane.

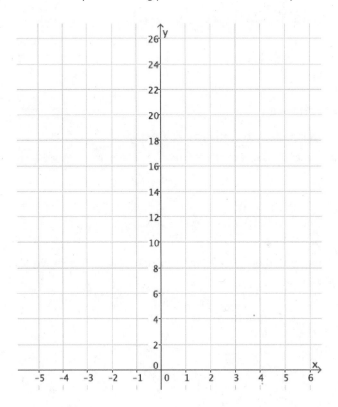

d.   What shape does the graph of the points appear to take?

e.   Find the rate of change using rows 1 and 2 from the table above.

f.   Find the rate of change using rows 2 and 3 from the table above.

EUREKA
MATH™

g. Find the rate of change using any two other rows from the table above.

h. Return to your initial claim about the function. Is it linear or nonlinear? Justify your answer with as many pieces of evidence as possible.

2. Consider the function that assigns to a number x the value x^3.

   a. Do you think the function is linear or nonlinear? Explain.

   b. Develop a list of inputs and outputs for this function. Organize your work using the table below. Then, answer the questions that follow.

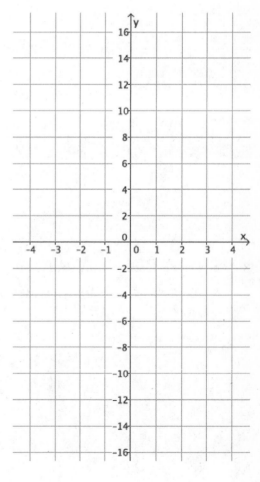

Input (x)	Output (x^3)
−2.5	
−2	
−1.5	
−1	
−0.5	
0	
0.5	
1	
1.5	
2	
2.5	

   c. Plot the inputs and outputs as ordered pairs defining points on the coordinate plane.

    d.   What shape does the graph of the points appear to take?

    e.   Find the rate of change using rows 2 and 3 from the table above.

    f.   Find the rate of change using rows 3 and 4 from the table above.

    g.   Find the rate of change using rows 8 and 9 from the table above.

    h.   Return to your initial claim about the function. Is it linear or nonlinear? Justify your answer with as many pieces of evidence as possible.

3.   Consider the function that assigns to each positive number x the value $\frac{1}{x}$.

    a.   Do you think the function is linear or nonlinear? Explain.

b.   Develop a list of inputs and outputs for this function. Organize your work using the table below. Then, answer the questions that follow.

Input (x)	Output $\left(\frac{1}{x}\right)$
0.1	
0.2	
0.4	
0.5	
0.8	
1	
1.6	
2	
2.5	
4	
5	

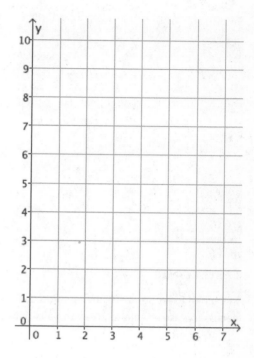

c.   Plot the inputs and outputs as ordered pairs defining points on the coordinate plane.

d.   What shape does the graph of the points appear to take?

e.   Find the rate of change using rows 1 and 2 from the table above.

f.   Find the rate of change using rows 2 and 3 from the table above.

g.   Find the rate of change using any two other rows from the table above.

h.   Return to your initial claim about the function. Is it linear or nonlinear? Justify your answer with as many pieces of evidence as possible.

Exercises 4–10

In each of Exercises 4–10, an equation describing a rule for a function is given, and a question is asked about it. If necessary, use a table to organize pairs of inputs and outputs, and then plot each on a coordinate plane to help answer the question.

4.   What shape do you expect the graph of the function described by $y = x$ to take? Is it a linear or nonlinear function?

5.   What shape do you expect the graph of the function described by $y = 2x^2 - x$ to take? Is it a linear or nonlinear function?

6.   What shape do you expect the graph of the function described by $3x + 7y = 8$ to take? Is it a linear or nonlinear function?

7. What shape do you expect the graph of the function described by $y = 4x^3$ to take? Is it a linear or nonlinear function?

8. What shape do you expect the graph of the function described by $\frac{3}{x} = y$ to take? Is it a linear or nonlinear function? (Assume that an input of $x = 0$ is disallowed.)

9. What shape do you expect the graph of the function described by $\frac{4}{x^2} = y$ to take? Is it a linear or nonlinear function? (Assume that an input of $x = 0$ is disallowed.)

10. What shape do you expect the graph of the equation $x^2 + y^2 = 36$ to take? Is it a linear or nonlinear function? Is it a function? Explain.

Lesson Summary

One way to determine if a function is linear or nonlinear is to inspect average rates of change using a table of values. If these average rates of change are not constant, then the function is not linear.

Another way is to examine the graph of the function. If all the points on the graph do not lie on a common line, then the function is not linear.

If a function is described by an equation different from one equivalent to $y = mx + b$ for some fixed values m and b, then the function is not linear.

Problem Set

1. Consider the function that assigns to each number x the value $x^2 - 4$.

    a. Do you think the function is linear or nonlinear? Explain.

    b. Do you expect the graph of this function to be a straight line?

    c. Develop a list of inputs and matching outputs for this function. Use them to begin a graph of the function.

    d. Was your prediction to (b) correct?

Input (x)	Output ($x^2 - 4$)
-3	
-2	
-1	
0	
1	
2	
3	

2. Consider the function that assigns to each number x greater than -3 the value $\dfrac{1}{x+3}$.

    a. Is the function linear or nonlinear? Explain.

    b. Do you expect the graph of this function to be a straight line?

    c. Develop a list of inputs and matching outputs for this function. Use them to begin a graph of the function.

    d. Was your prediction to (b) correct?

Input (x)	Output $\left(\dfrac{1}{x+3}\right)$
-2	
-1	
0	
1	
2	
3	

3.

    a.   Is the function represented by this graph linear or nonlinear? Briefly justify your answer.

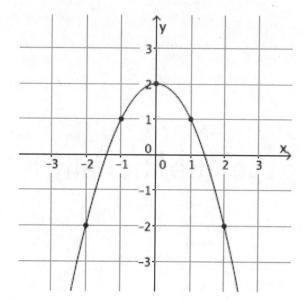

    b.   What is the average rate of change for this function from an input of $x = -2$ to an input of $x = -1$?

    c.   What is the average rate of change for this function from an input of $x = -1$ to an input of $x = 0$?

This page intentionally left blank

Lesson 9: Examples of Functions from Geometry

Classwork

Exploratory Challenge 1/Exercises 1–4

As you complete Exercises 1–4, record the information in the table below.

	Side length in inches (s)	Area in square inches (A)	Expression that describes area of border
Exercise 1			
Exercise 2			
Exercise 3			
Exercise 4			

1. Use the figure below to answer parts (a)–(f).

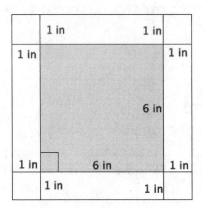

a. What is the length of one side of the smaller, inner square?

b. What is the area of the smaller, inner square?

c. What is the length of one side of the larger, outer square?

d. What is the area of the larger, outer square?

e. Use your answers in parts (b) and (d) to determine the area of the 1-inch white border of the figure.

f. Explain your strategy for finding the area of the white border.

EUREKA
MATH™

2. Use the figure below to answer parts (a)–(f).

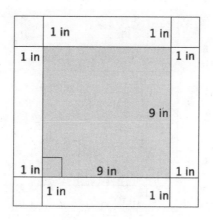

a. What is the length of one side of the smaller, inner square?

b. What is the area of the smaller, inner square?

c. What is the length of one side of the larger, outer square?

d. What is the area of the larger, outer square?

e. Use your answers in parts (b) and (d) to determine the area of the 1-inch white border of the figure.

f. Explain your strategy for finding the area of the white border.

EUREKA
MATH™

Lesson 9:   Examples of Functions from Geometry

© 2015 Great Minds. eureka-math.org
G8-M3M4M5-SE-B2-1.3.1-01.2016

S.65

3. Use the figure below to answer parts (a)–(f).

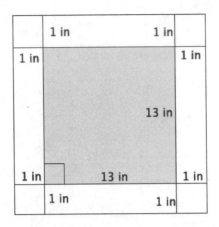

a. What is the length of one side of the smaller, inner square?

b. What is the area of the smaller, inner square?

c. What is the length of one side of the larger, outer square?

d. What is the area of the larger, outer square?

e. Use your answers in parts (b) and (d) to determine the area of the 1-inch white border of the figure.

f. Explain your strategy for finding the area of the white border.

© 2015 Great Minds. eureka-math.org
G8-M3M4M5-SE-B2-1.3.1-01.2016

4. Write a function that would allow you to calculate the area of a 1-inch white border for any sized square picture measured in inches.

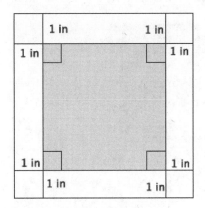

a. Write an expression that represents the side length of the smaller, inner square.

b. Write an expression that represents the area of the smaller, inner square.

c. Write an expression that represents the side lengths of the larger, outer square.

d. Write an expression that represents the area of the larger, outer square.

e. Use your expressions in parts (b) and (d) to write a function for the area A of the 1-inch white border for any sized square picture measured in inches.

Exercises 5–6

5. The volume of the prism shown below is 61.6 in³. What is the height of the prism?

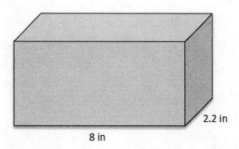

8 in

2.2 in

6. Find the value of the ratio that compares the volume of the larger prism to the smaller prism.

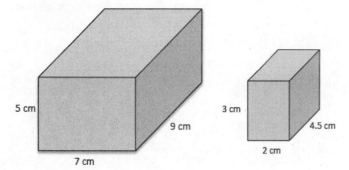

5 cm

9 cm

7 cm

3 cm

4.5 cm

2 cm

Lesson 9:   Examples of Functions from Geometry

EUREKA
MATH™

Exploratory Challenge 2/Exercises 7–10

As you complete Exercises 7–10, record the information in the table below. Note that *base* refers to the bottom of the prism.

	Area of base in square centimeters (B)	Height in centimeters (h)	Volume in cubic centimeters
Exercise 7			
Exercise 8			
Exercise 9			
Exercise 10			

7. Use the figure to the right to answer parts (a)–(c).

    a. What is the area of the base?

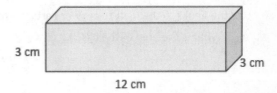

3 cm
12 cm
3 cm

    b. What is the height of the figure?

    c. What is the volume of the figure?

8. Use the figure to the right to answer parts (a)–(c).

    a. What is the area of the base?

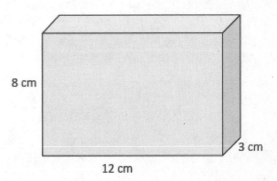

8 cm
3 cm
12 cm

    b. What is the height of the figure?

    c. What is the volume of the figure?

9. Use the figure to the right to answer parts (a)–(c).

    a.   What is the area of the base?

    b.   What is the height of the figure?

    c.   What is the volume of the figure?

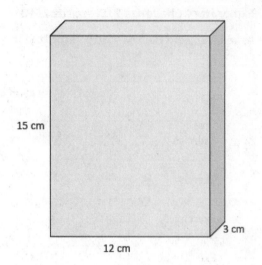

15 cm

3 cm

12 cm

10. Use the figure to the right to answer parts (a)–(c).

    a.   What is the area of the base?

    b.   What is the height of the figure?

    c.   Write and describe a function that will allow you to
         determine the volume of any rectangular prism that has a
         base area of 36 cm^2.

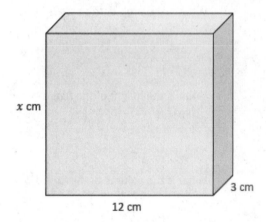

x cm

3 cm

12 cm

Lesson 9:   Examples of Functions from Geometry

**EUREKA
MATH**™

Lesson Summary

There are a few basic assumptions that are made when working with volume:

(a)   The volume of a solid is always a number greater than or equal to 0.

(b)   The volume of a unit cube (i.e., a rectangular prism whose edges all have a length of 1) is by definition 1 cubic unit.

(c)   If two solids are identical, then their volumes are equal.

(d)   If two solids have (at most) their boundaries in common, then their total volume can be calculated by adding the individual volumes together. (These figures are sometimes referred to as composite solids.)

Problem Set

1.   Calculate the area of the 3-inch white border of the square figure below.

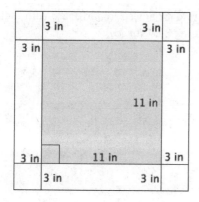

2.   Write a function that would allow you to calculate the area, A, of a 3-inch white border for any sized square picture measured in inches.

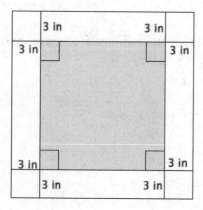

3. Dartboards typically have an outer ring of numbers that represent the number of points a player can score for getting a dart in that section. A simplified dartboard is shown below. The center of the circle is point A. Calculate the area of the outer ring. Write an exact answer that uses π (*do not* approximate your answer by using 3.14 for π).

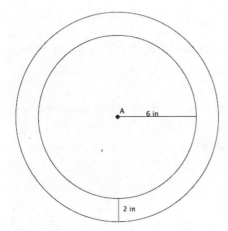

4. Write a function that would allow you to calculate the area, A, of the outer ring for any sized dartboard with radius r. Write an exact answer that uses π (*do not* approximate your answer by using 3.14 for π).

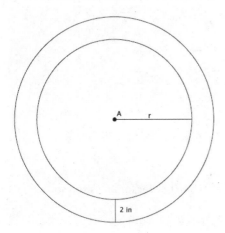

5. The shell of the solid shown was filled with water and then poured into the standard rectangular prism, as shown. The height that the volume reaches is 14.2 in. What is the volume of the shell of the solid?

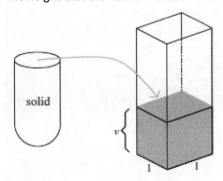

EUREKA
MATH

6. Determine the volume of the rectangular prism shown below.

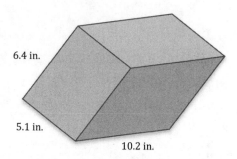

6.4 in.

5.1 in.

10.2 in.

7. The volume of the prism shown below is 972 cm³. What is its length?

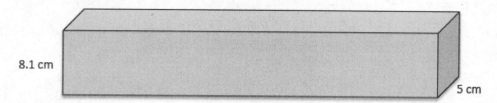

8.1 cm

5 cm

8. The volume of the prism shown below is 32.7375 ft³. What is its width?

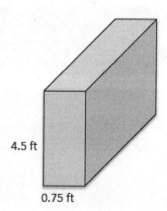

4.5 ft

0.75 ft

9. Determine the volume of the 3-dimensional figure below. Explain how you got your answer.

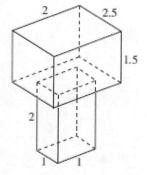

2   2.5

1.5

2

1   1

EUREKA
MATH™

Lesson 9:   Examples of Functions from Geometry

S.73

© 2015 Great Minds. eureka-math.org
G8-M3M4M5-SE-B2-1.3.1-01.2016

This page intentionally left blank

Lesson 10: Volumes of Familiar Solids—Cones and Cylinders

Classwork

Opening Exercise

    a.

        i.   Write an equation to determine the volume of the rectangular prism shown below.

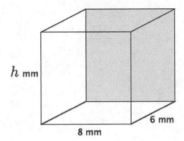

        ii.   Write an equation to determine the volume of the rectangular prism shown below.

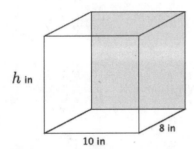

        iii.   Write an equation to determine the volume of the rectangular prism shown below.

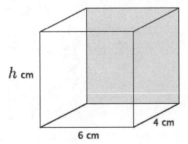

iv. Write an equation for volume, V, in terms of the area of the base, B.

b. Using what you learned in part (a), write an equation to determine the volume of the cylinder shown below.

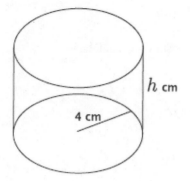

Exercises 1–3

1. Use the diagram to the right to answer the questions.

a. What is the area of the base?

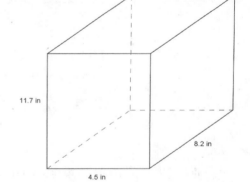

b. What is the height?

c. What is the volume of the rectangular prism?

2. Use the diagram to the right to answer the questions.

   a. What is the area of the base?

   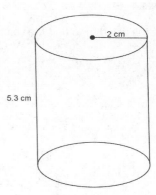

   b. What is the height?

   c. What is the volume of the right circular cylinder?

3. Use the diagram to the right to answer the questions.

   a. What is the area of the base?

   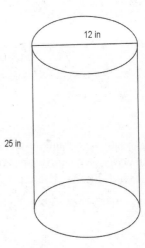

   b. What is the height?

   c. What is the volume of the right circular cylinder?

Exercises 4–6

4. Use the diagram to find the volume of the right circular cone.

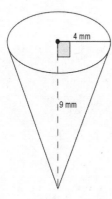

4 mm

9 mm

5. Use the diagram to find the volume of the right circular cone.

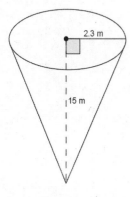

2.3 m

15 m

6. Challenge: A container in the shape of a right circular cone has height h, and base of radius r, as shown. It is filled with water (in its upright position) to half the height. Assume that the surface of the water is parallel to the base of the inverted cone. Use the diagram to answer the following questions:

a. What do we know about the lengths of AB and AO?

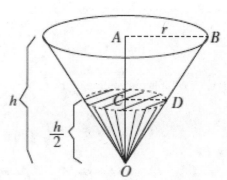

b.   What do we know about the measure of $\angle OAB$ and $\angle OCD$?

c.   What can you say about $\triangle OAB$ and $\triangle OCD$?

d.   What is the ratio of the volume of water to the volume of the container itself?

Lesson Summary

The formula to find the volume, V, of a right circular cylinder is $V = \pi r^2 h = Bh$, where B is the area of the base.

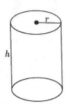

The formula to find the volume of a cone is directly related to that of the cylinder. Given a right circular cylinder with radius r and height h, the volume of a cone with those same dimensions is one-third of the cylinder. The formula for the volume, V, of a circular cone is $V = \frac{1}{3}\pi r^2 h$. More generally, the volume formula for a general cone is $V = \frac{1}{3}Bh$, where B is the area of the base.

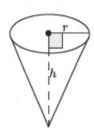

Problem Set

1. Use the diagram to help you find the volume of the right circular cylinder.

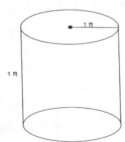

EUREKA MATH™

2. Use the diagram to help you find the volume of the right circular cone.

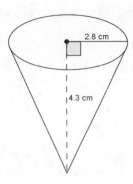

2.8 cm

4.3 cm

3. Use the diagram to help you find the volume of the right circular cylinder.

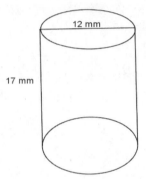

12 mm

17 mm

4. Use the diagram to help you find the volume of the right circular cone.

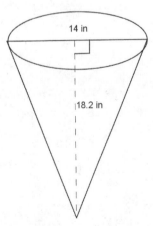

14 in

18.2 in

5. Oscar wants to fill with water a bucket that is the shape of a right circular cylinder. It has a 6-inch radius and 12-inch height. He uses a shovel that has the shape of a right circular cone with a 3-inch radius and 4-inch height. How many shovelfuls will it take Oscar to fill the bucket up level with the top?

6. A cylindrical tank (with dimensions shown below) contains water that is 1-foot deep. If water is poured into the tank at a constant rate of 20 $\frac{ft^3}{min}$ for 20 min., will the tank overflow? Use 3.14 to estimate π.

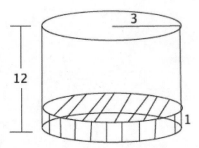

Volumes of Familiar Solids—Cones and Cylinders

© 2015 Great Minds. eureka-math.org
G8-M3M4M5-SE-B2-1.3.1-01.2016

EUREKA
MATH™

Lesson 11: Volume of a Sphere

Classwork

Exercises 1–3

1. What is the volume of a cylinder?

2. What is the height of the cylinder?

3. If volume(sphere) $= \frac{2}{3}$ volume(cylinder with same diameter and height), what is the formula for the volume of a sphere?

Example 1

Compute the exact volume for the sphere shown below.

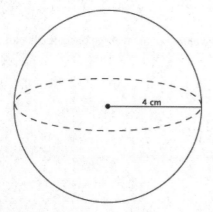

Example 2

A cylinder has a diameter of 16 inches and a height of 14 inches. What is the volume of the largest sphere that will fit into the cylinder?

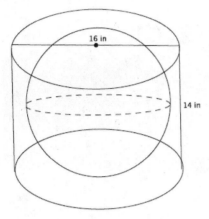

Exercises 4–8

4. Use the diagram and the general formula to find the volume of the sphere.

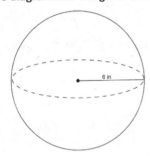

5. The average basketball has a diameter of 9.5 inches. What is the volume of an average basketball? Round your answer to the tenths place.

EUREKA
MATH™

6. A spherical fish tank has a radius of 8 inches. Assuming the entire tank could be filled with water, what would the volume of the tank be? Round your answer to the tenths place.

7. Use the diagram to answer the questions.

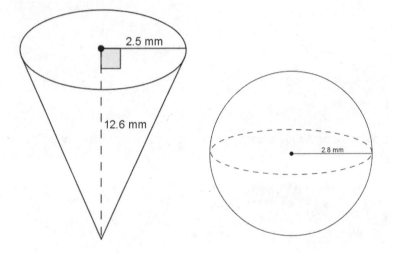

2.5 mm

12.6 mm

2.8 mm

a. Predict which of the figures shown above has the greater volume. Explain.

b.   Use the diagram to find the volume of each, and determine which has the greater volume.

8.   One of two half spheres formed by a plane through the sphere's center is called a hemisphere. What is the formula for the volume of a hemisphere?

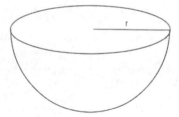

EUREKA
MATH™

Lesson Summary

The formula to find the volume of a sphere is directly related to that of the right circular cylinder. Given a right circular cylinder with radius r and height h, which is equal to $2r$, a sphere with the same radius r has a volume that is exactly two-thirds of the cylinder.

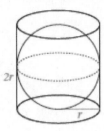

Therefore, the volume of a sphere with radius r has a volume given by the formula $V = \frac{4}{3}\pi r^3$.

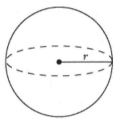

Problem Set

1. Use the diagram to find the volume of the sphere.

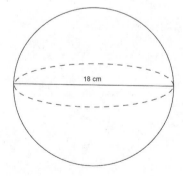

18 cm

2.   Determine the volume of a sphere with diameter 9 mm, shown below.

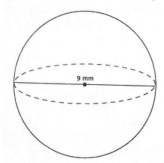

3.   Determine the volume of a sphere with diameter 22 in., shown below.

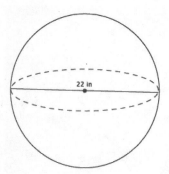

4.   Which of the two figures below has the lesser volume?

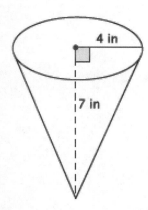

5. Which of the two figures below has the greater volume?

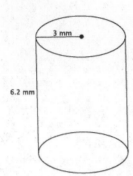
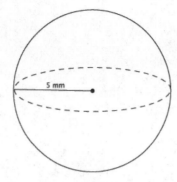

6. Bridget wants to determine which ice cream option is the best choice. The chart below gives the description and prices for her options. Use the space below each item to record your findings.

$2.00	$3.00	$4.00
One scoop in a cup	Two scoops in a cup	Three scoops in a cup
Half a scoop on a cone filled with ice cream		A cup filled with ice cream (level to the top of the cup)

A scoop of ice cream is considered a perfect sphere and has a 2-inch diameter. A cone has a 2-inch diameter and a height of 4.5 inches. A cup, considered a right circular cylinder, has a 3-inch diameter and a height of 2 inches.

a. Determine the volume of each choice. Use 3.14 to approximate π.

b. Determine which choice is the best value for her money. Explain your reasoning.

EUREKA
MATH™

Lesson 11:   Volume of a Sphere

S.89

© 2015 Great Minds. eureka-math.org
G8-M3M4M5-SE-B2-1.3.1-01.2016

This page intentionally left blank